recent advances in phytochemistry

volume 13

Topics in the Biochemistry of Natural Products

RECENT ADVANCES IN PHYTOCHEMISTRY

Recent Volumes in the Series

Volume 9 **Phytochemistry in Disease and Medicine**
Proceedings of the Fourteenth Annual Meeting, Western Carolina University, Cullowhee, North Carolina, August, 1974

Volume 10 **Biochemical Interaction Between Plants and Insects**
Proceedings of the Fifteenth Annual Meeting, Tampa, Florida, August, 1975

Volume 11 **The Structure, Biosynthesis, and Degradation of Wood**
Proceedings of the Sixteenth Annual Meeting, Vancouver, Canada, August, 1976

Volume 12 **Biochemistry of Plant Phenolics**
Proceedings of the Joint Symposium of the Phytochemical Society of Europe and the Phytochemical Society of North America, Ghent, Belgium, August, 1977

Volume 13 **Topics in the Biochemistry of Natural Products**
Proceedings of the First Joint Meeting of the American Society of Pharmacognosy and the Phytochemical Society of North America, Stillwater, Oklahoma, August, 1978

---

recent advances in phytochemistry

volume 13

Topics in the Biochemistry of Natural Products

Edited by
Tony Swain
Boston University
Boston, Massachusetts

and
George R. Waller
Oklahoma State University
Stillwater, Oklahoma

PLENUM PRESS • NEW YORK AND LONDON

Library of Congress Cataloging in Publication Data

Main entry under title:

Topics in the biochemistry of natural products.

(Recent advances in phytochemistry; v. 13)
"Proceedings of the first joint meeting of the American Society of Pharmacognosy and the Phytochemical Society of North America, held at Oklahoma State University, Stillwater, Oklahoma, August 14–17, 1978."
Includes index.
1. Enzymes—Congresses. 2. Botanical chemistry—Congresses. 3. Biological products—Congresses. I. Swain, T. II. Waller, G. R. III. American Society of Pharmacognosy. IV. Phytochemical Society of North America. V. Series.
QK861.R38 vol. 13 [QK898.P58] 581.1'9'280s [581.1'924]
ISBN 0-306-40188-6 79-16473

Proceedings of the First Joint Meeting of the American Society of Pharmacognosy and the Phytochemical Society of North America, held at Oklahoma State University, Stillwater, Oklahoma, August 14–17, 1978

A Division of Plenum Publishing Corporation
227 West 17th Street, New York, N.Y. 10011

Printed in the United States of America

Contributors

D. J. ABERHART, Worcester Foundation for Experimental Biology, Shrewsbury, Massachusetts

ROBERT ANTON, Laboratoire de Matiere Medicale, Universite Louis Pasteur, Strasbourg, France

CARMINE J. COSCIA, Department of Biochemistry, Saint Louis University School of Medicine, St. Louis, Missouri

MARK W. DUDLEY, Division of Biochemistry, University of California, Los Angeles, California

MICHAEL T. DUEBER, Division of Biochemistry, University of California, Los Angeles, California

WILLIAM FENICAL, Institute of Marine Resources, Scripps Institution of Oceanography, La Jolla, California

K. M. MADYASTHA, Department of Biochemistry, Saint Louis University School of Medicine, St. Louis, Missouri

GUY OURISSON, Institute de Chimie, Universite Louis Pasteur, Strasbourg, France

RONALD J. PARRY, Department of Chemistry, Rice University, Houston, Texas

JANOS RETEY, Lehrstuhl fur Biochemie, Universitat Karlsruhe, Karlsruhe, West Germany

MICHEL ROHMER, Institut de Chimie, Universite Louis Pasteur, Strasbourg, France

YUZURU SHIMIZU, Department of Pharmacognosy, University of Rhode Island, Kingston, Rhode Island

CHARLES A. WEST, Division of Biochemistry, University of California, Los Angeles, California

Preface

The study of plant natural products can be considered to have started in 1806 when Serturner isolated the first of these compounds, morphine, from the opium poppy. Over the next 150 years, numerous elegant and powerful techniques were developed for the isolation and structural identification of a variety of classes of these substances. With increasing knowledge, various speculations were put forward on the ways that these compounds were synthesised from primary metabolites and on their possible physiological, ecological and taxonomic importance.

These investigations sublimated in the late 1940's with the newly developed application of radioactively labelled precursors to the study various biosynthetic pathways and use of rapid chromatagraphic and spectroscopic methods to uncover the structure of a plethora of new compounds and to recognize their wide distribution in various taxa of plants and other organisms.

We have now entered a new stage in our investigation of natural products in which several important new problems are being tackled. The present emphasis is on how the various biosynthetic pathways are regulated, the enzymic parameters which determine the chirality of the products, the transfer and modification of components in food chains, and the ways in which a given class of physiologically or ecologically important compounds has been superseded by another during the course of evolution. All these problems were discussed by the distinguished international contributors to the Joint Symposium of the Phytochemical Society of North America and the American Society of Pharmacognasy which was held in August 1978 on the campus of Oklahoma State University at Stillwater. Their presentations form the eight chapters of this volume. Although the main emphasis is on alkaloids, amino acids and terpenoids, reference is made to most other classes of naturally occurring substances, and a number of general principles are outlined.

The Symposium, which was superbly organised by my co-editor, Professor George Waller and his colleagues, included two important contributions on science policy. These were given by Oklahoma Senator, the Honorable Henry Bellman and Dr. Eloise Clark of the National Science Foundation. On behalf of the two Societies, I wish to thank both of them for sharing their thought provoking views on science support with the participants of the joint symposium.

These days, no international meeting can be held without a geat deal of outside financial support. The Societies were fortunate to obtain grants to cover over one third of the cost of the Symposium from Beckman Analytical Instruments, Cooper Laboratories Inc., Johnson and Johnson, Kerr-McGee Foundation, Thomas J. Lipton Inc., McNeil Laboratories, The Merrick Foundation, Miles Laboratories Inc., Oklahoma State University, Pennwalt Corporation, Public Service Co. of Oklahoma, Searle Laboratories, and The Upjohn Company. It is my pleasure on behalf of the PNSA and the ASP, to thank all these Companies and Institutions for their generous support.

Finally, I am grateful to the authors for their prompt delivery of manuscripts and to Plenum Press for their expert production of this volume.

T. Swain
March 1979

Contents

MARINE NATURAL PRODUCTS

Chapter One

ARE THE STERIC COURSES OF ENZYMATIC REACTIONS INFORMATIVE ABOUT THEIR MECHANISMS?

JANOS RETEY

Lehrstuhl für Biochemie
im Institut fur Organische Chemie
Universitat Karlsruhe
7500 Karlsruhe

INTRODUCTION

Most of the biologically active substances: drugs, hormones, vitamins, flavors etc. occur as only one of several possible stereoisomers. Nature makes them in a completely stereospecific manner. In contrast, organic chemists have great difficulty in finding highly stereospecific reactions when synthesizing such compounds. It is not surprising therefore that stereospecificity is playing an increasingly important role both in the synthesis of natural products and in our understanding of reaction mechanisms. When chemists convert one product into another and cannot demonstrate the existence of stable intermediates in a direct way, we may talk of a black box process. We can of course still ask, what happens in the black box -- what happens between the starting material and the product?

The intermediate structure of highest energy, the transition state, cannot be observed directly by definition. Experimental access to the transition state is possible in two ways. A kinetic analysis allows an estimation of the activation energy, while a determination of the steric course of a reaction allows conclusions to be drawn as to the transient structural changes. If stereochemical purity is conserved, one can construct transition state structures which are stabilized by orbital interactions and simultaneously explain the observed stereochemical course of the reaction. A classic example for this is the Walden inversion in S_N2-type substitutions, while a more recent correlation between the steric course and the transition state structure is provided by the Woodward-Hoffmann rules. If, however, stereochemical purity is lost during a reaction, some metastable intermediate must occur which might for example proffer geometrically equivalent faces (Figure 1 A and B) to attack by a reagent or which might change its geometry or handedness (sense of chirality) by rapid processes, such as rotation around single bonds (Figure 1 C) or inversion (Figure 1 D). In pure chemical reactions therefore, both conservation and loss of stereochemical purity can provide useful information about the mechanism.

Since Pasteur's time, enzymatic reactions have been known to exhibit a very high degree of stereospecificity and it was only natural to believe that the determination of their steric course would shed light on their mechanism. Already at this stage, however, it should be pointed out that the term enzymatic stereospecificity covers two distinct phenomena. First, it includes the overall steric course of the reaction, as applied to pure chemical processes, e.g. substitution reactions or addition-elimination reactions involving double bonds, and second it includes chiral recognition. One aspect of chiral recognition, namely the ability to distinguish between reflection-equivalent structures (i.e. enantiomers), was recognized long ago by Pasteur. An appreciation of a second more subtle aspect, namely the ability to distinguish between reflection-equivalent partial structures (enantiotopic groups or faces) within a molecule, had to wait until 1948, when it was pointed out by Ogston.

Experimental observation of this second aspect of chiral recognition requires in many cases stereospecific isotopic labeling of one of the enantiotopic groups. The question as

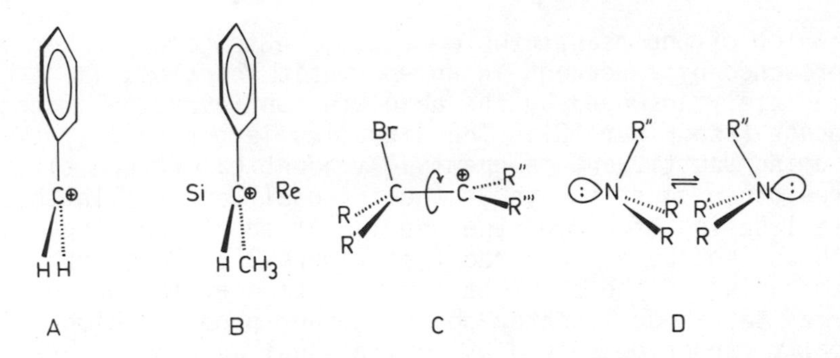

Figure 1. Metastable intermediates with geometrically equivalent (A stereohomotopic, B enantiotopic) trigonal faces, with the possibility of rotation around a single bond (C) and of inversion (D).

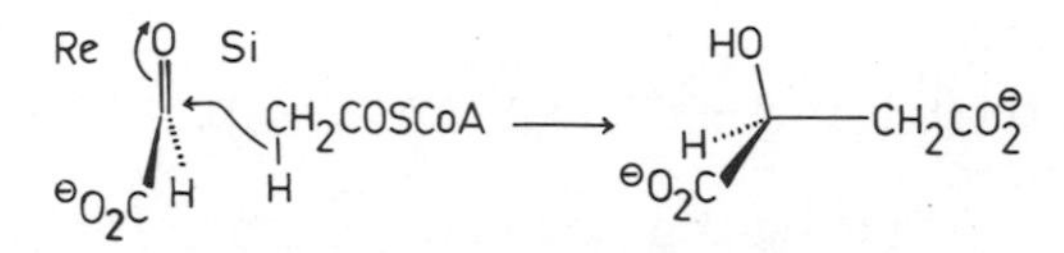

Figure 2. Attack on the Si-face of the carbonyl group of glyoxylate by acetyl-CoA in the malate synthase reaction. From the (2S)-configuration of the product one can follow immediately to the stereospecificity of the attack.

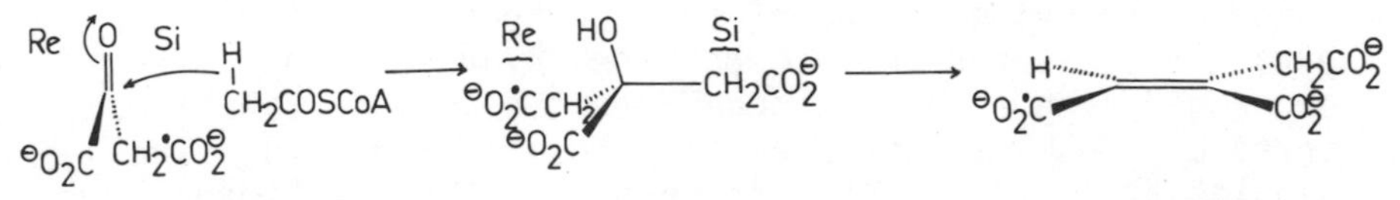

Figure 3. Attack on the Si-face of the carbonyl group of oxalacetate by acetyl-CoA in the citrate synthase reaction. The stereospecificity of the attack can only be observed by isotopic labeling (e.g. $^{\circ}C$-^{14}C). Aconitase abstracts water from the Re-branch of citrate to form cis-aconitate.

to which of the enantiotopic faces of a trigonal center is approached by a reagent in an enzymatic reaction, is often immediately answered by the absolute configuration of the product (see Figure 2). The situation is not so easy if the incoming substituent is chemically identical to one of the three other substituents at the trigonal center. In this case labeling of one of the chemically equal substituents with an isotope is required (see Figure 3). Moreover, determination of the absolute configuration at the newly formed tetrahedral center poses another problem which usually cannot be solved by conventional methods. Advantageously, a second enzymatic system with known stereospecificity may be used for such a purpose, e.g. in the case of asymmetrically labeled citrate, this second enzyme may be aconitase, for which it has been shown, by using labeled substrate of known absolute configuration, that attack occurs specifically at the $(CH_2COO^-)_{\underline{Re}}$ branch of citrate (Figure 3 [1,2]).

CHIRAL RECOGNITION BY PHENYLPYRUVATE TAUTOMERASE

Stereospecific isotopic labeling of enantiotopic groups can be achieved either chemically or enzymatically. A recent example of the enzymatic introduction of deuterium into an enantiotopic position is illustrated in Figure 4 [3]. The commercially available enzyme, beef liver tautomerase, catalyzes the enolization of phenylpyruvate. As can be shown by ^1H-NMR spectroscopy in deuterium oxide, about 2-3% of the enol form is present at equilibrium. The rate of spontaneous tautomerization is rather slow at pH 6.2 but is increased several hundred fold by the enzyme. At each tautomerization step, one solvent proton will be incorporated into the methylene group of the substrate. If the incorporation is stereospecific only one hydrogen atom will be exchanged even after prolonged reaction. This can be monitored by ^1H-NMR spectroscopy as shown in Figure 5. The singlet at δ = 4.12 ppm corresponds to the enantiotopic methylene protons in unlabeled phenylpyruvate and the incompletely resolved triplet to that of monodeuterated phenylpyruvate. Under the experimental conditions one deuterium atom is incorporated after 40 min and the spectrum does not then significantly change during the next 13 hours. Supportive evidence for stereospecific exchange was provided by the circular dichroism spectrum of the monodeuterated phe-

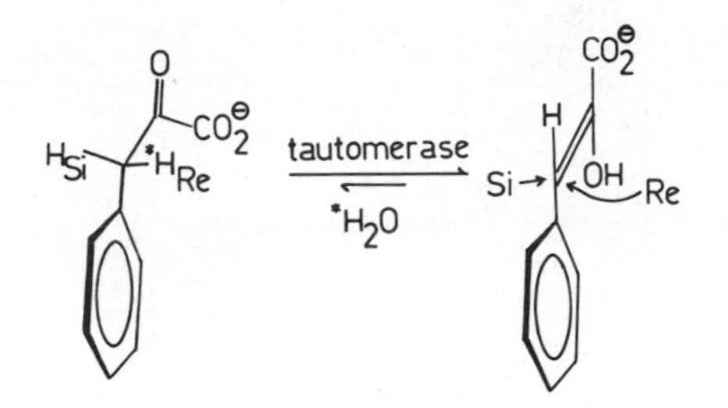

Figure 4. The stereospecificity of tautomerase. The enzmye catalyzes the reversible addition of a proton to the Re-face of phenyl-enolpyruvate.

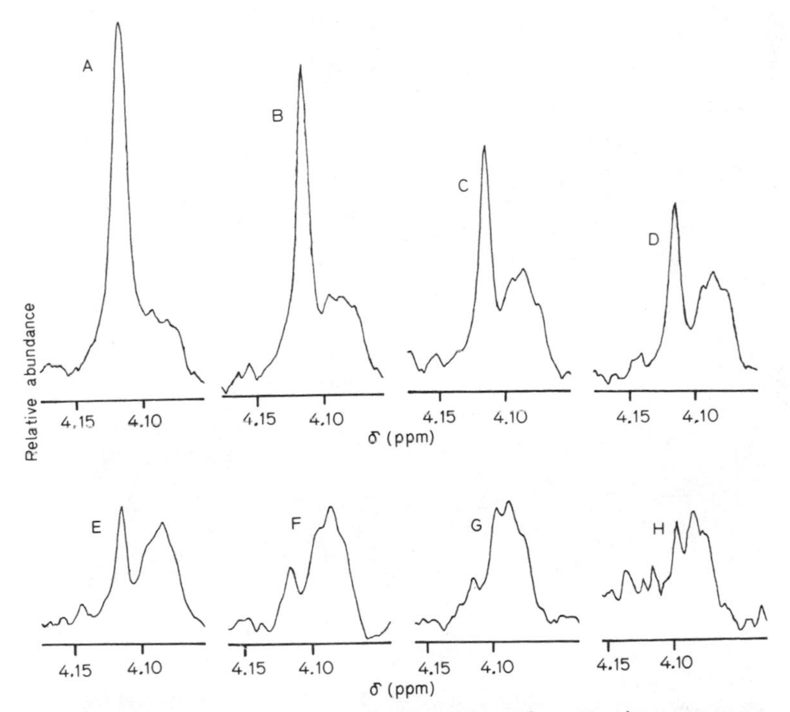

Figure 5. Kinetics of the deuterium-exchange reaction in phenylpyruvate (4.9 M) catalyzed by tautomerase in deuterium oxide. A part of the 270 MHz ^{1}H-NMR spectra around 4 ppm 4 min (A), 6.5 min (B), 9 min (C), 11.4 min (D), 14 min (E), 20 min (F), 41.7 min (G) and 13 h 40 min (H) after start of the reaction. (Taken from ref. 3 with permission of Eur. J. Biochem.)

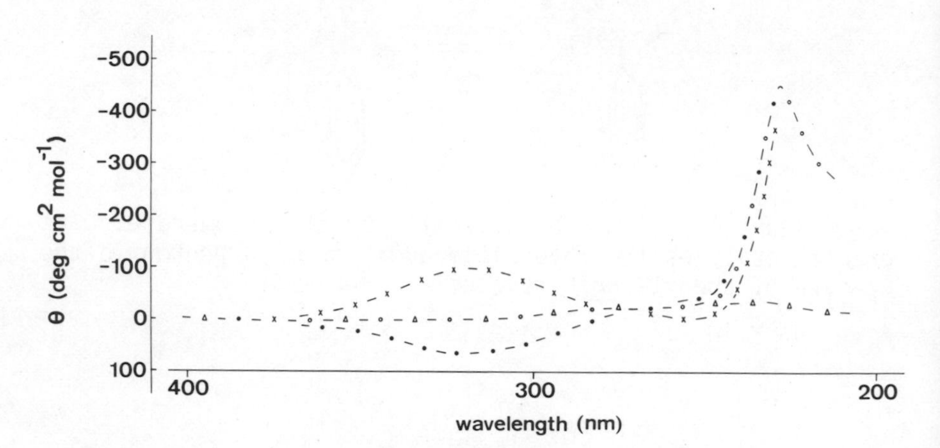

Figure 6. Circular dichroism spectra of (3_R_)-3-phenyl-[3-$^2H_1$]-pyruvate + tautomerase (I); (3_S_)-3-phenyl-[3-$^2H_1$]-pyruvate + tautomerase (II), tautomerase (III) and water (IV). (Taken from ref. 3 with permission of Eur. J. Biochem.)

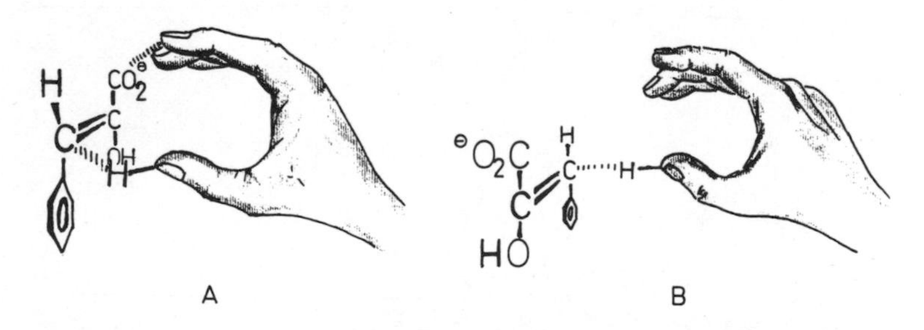

Figure 7. Demonstration of the geometrically different (diastereomeric) relationship between a chiral object and the enantiotopic trigonal faces of 3-phenyl-_enol_-pyruvate. Approach to the _Re_-face (A) and approach to the _Si_-face (B).

nylpyruvate. After a few hours incubation of the substrate with tautomerase in deuterium oxide, a negative Cotton effect at 319 nm could be observed (Figure 6). The enantiomeric (3S)-[3-$^2H_1$]-phenylpyruvate can be obtained in an analogous manner by incubation of [3-$^2H_2$]-phenylpyruvate with the tautomerase in normal water. Comparison of the behavior of a reference [$^2H_1$]-phenylpyruvate of known absolute configuration (4) revealed that tautomerase removes stereospecifically the 3-H_{Re}-proton of the substrate. According to the principle of microscopic reversibility, tautomerase adds a proton in the reverse reaction to the Re-face of enol-phenylpyruvate. Indeed an approach by the enzyme-bound proton from the Si-face would lead to a diastereomeric relationship and consequently to a diastereomeric transition state of apparently much higher energy (Figure 7).

The tautomerase reaction thus provides a good example of the chiral recognition of reflection-equivalent (enantiotopic) partial structures and also illustrates well the power of stereospecific labeling of enantiotopic groups. Chemical labeling of the enantiotopic protons would be hampered by the spontaneous and therefore non-stereospecific enolization of the substrate. Phenylpyruvate and 4-hydroxyphenylpyruvate are in equilibrium with phenylalanine and tyrosine, respectively, through the action of aminotransferases present in most cells. They could therefore be used as stereospecifically labeled precursors in biosynthetic studies.

Chiral recognition of this type does not provide mechanistically relevant information about the reaction until the three-dimensional structure of the enzyme is known. From our experience with chemical systems we can see that the interaction between an enzyme and its substrate must be very intimate and unambiguous in order to account for such a high degree of stereoselectivity.

THE STEREOSPECIFICITY OF COENZYME-B_{12}-CATALYZED REARRANGEMENTS (GENERAL)

One would assume that the steric course of the reaction, especially in enzymatic processes involving substitutions, would allow us to draw useful conclusions as to the

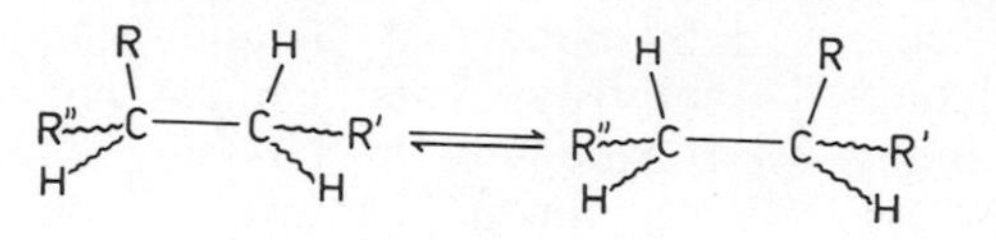

R = COSCoA, R' = CO_2H, R'' = H, CH_3; methylmalonyl CoA mutase

R = $CH(NH_2)CO_2H$, R' = CO_2H, R'' = H; glutamate mutase

R = $C(=CH_2)CO_2H$, R' = CO_2H, R'' = H; 2-methylene glutarate mutase

R = R' = OH, R'' = CH_3, H, CH_2OH; diol dehydrase

R = NH_2, R' = OH, R'' = H, CH_3; ethanolamine ammonia lyase

R = NH_2, R' = $CH_2CH(NH_2)CH_2CO_2H$, R'' = H; ß-lysine mutase

Figure 8. Coenzyme-B_{12}-catalyzed rearrangements.

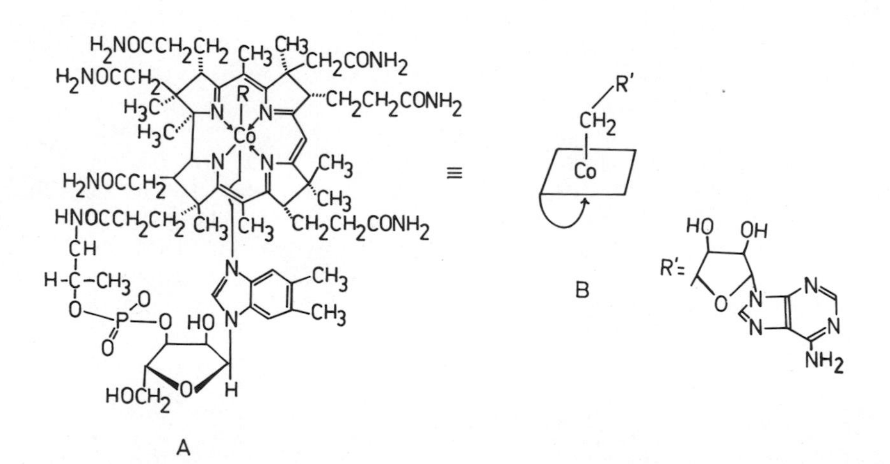

Figure 9. The structure of coenzyme B_{12} (A) and its abbreviated formula (B).

mechanism in the same way as in pure chemical reactions. The rearrangements catalyzed by coenzyme-B_{12}-dependent enzymes* seemed at the outset to be very promising objectives for stereochemical studies because (a) two substitutions occur at adjacent carbon centers in each reaction and (b) experimental results relevant to the mechanism were badly needed, since no analogy for these rearrangements existed in organic chemistry. As shown in Figure 8 a hydrogen atom and a variable group (R) are interchanged between two adjacent carbon atoms. Whereas the migration of R has been shown to take place intramolecularly [6,7], experiments with tritiated substrates indicated[8-10] a transient transfer of the migrating hydrogen atom to the cobalt-bound methylene group (i.e. the 5'-position) of the coenzyme (for structure see Figure 9).

These and other results suggest that coenzyme B_{12} is intimately involved in the catalysis. The role of the enzyme protein is rather obscure, but it seems to be responsible for the substrate specificity, and as we shall see, most likely for the stereospecificity as well.

The early results on the steric course of coenzyme-B_{12}-catalyzed rearrangements have been summarized in several review articles[11-15] and a short recapitulation should suffice here. Inversion of configuration has been observed in the reactions with (2<u>R</u>)- and (2<u>S</u>)-propane-1,2-diol (I, II)[16,17], glycerol (III)[18], (2<u>S</u>, 3<u>S</u>)-methylaspartate (IV)[19] and (3<u>S</u>)-β-lysine (V)[20]. In contrast, the substitution takes place with retention in the reaction catalyzed by methylmalonyl-CoA mutase (see VI in Figure 10)[21]. In each case the steric course has been determined only at one of the migration termini (labeled with asterisks) and not at the other (labeled with filled circles). In the case of (2<u>R</u>)- and (2<u>S</u>)-propane-1,2-diol (I and II)[22] and in that of glycerol (III)[18] evidence has also been obtained for a stereospecific substitution at the second migration center, as revealed by the stereospecific dehydration of the inter-

* Until recently it might have been regarded as inappropriate to devote a considerable portion of a chapter appearing in Recent Advances of Phytochemistry to the stereospecificity of coenzyme-B_{12}-catalyzed reactions. Luckily enough, coenzyme-B_{12}-dependent enzymes have recently been detected also in plants[5].

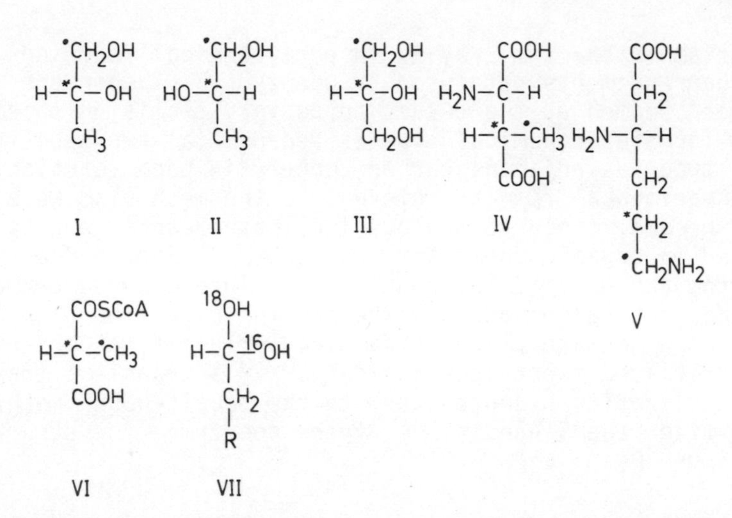

Figure 10. Substrates and intermediates of coenzyme-B_{12}-catalyzed rearrangements. The two migration termini are marked with filled circles and stars, respectively. (2R)-Propane-1,2-diol (I), (2S)-propane-1,2-diol (II), glycerol (III), (2S, 3S)-methylaspartate (IV), β-lysine (V), (2R)-methylmalonyl-CoA (VI), [1-$^{18}O_1$]-propane-1,1-diol (VII).

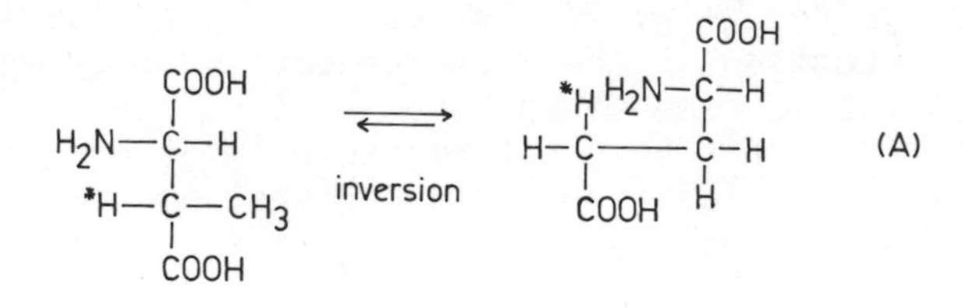

Figure 11. Steric course of the glutamate mutase (A) and methylmalonyl-CoA mutase (B) reactions as revealed by deuterium labeling.

mediate geminal diol (VII). However, elucidation of the stereochemical nature (retention or inversion) of the substitution was hampered by the impossibility of determining the sense of chirality of the intermediate geminal diol when asymmetrically labeled with oxygen isotopes (VII).

At this stage, the stereochemical results are more confusing than useful for mechanistic interpretation. For instance, the opposite steric course in the glutamate mutase and the methylmalonyl-CoA mutase reactions (Figure 11) would suggest that these two processes have completely different mechanisms. On the other hand the strikingly similar structural features of the two substrates, the almost identical rearrangement patterns, and the common coenzyme involved make such a conclusion unlikely.

Independently of the stereochemical results, data arising from isotope labeling[8,9,10,22], electron paramagnetic resonance [23-26] and UV/VIS-spectroscopic measurements[27] led to the working hypothesis for the mechanism of coenzyme-B_{12}-catalyzed rearrangements sketched in Figure 12[11]. In this scheme intermediates are postulated in which the migration termini are trigonal carbon centers and it is left open which of the stereoheterotopic faces of these centers will be approached by the attacking substituent. All the migration centers at which the steric course has been elucidated (labeled with asterisks in Figure 10) will be transformed, according to this hypothetical mechanism, into trigonal centers exhibiting either diastereotopic or enantiotopic faces. According to the same mechanism, some of the migration centers at which the steric course has not yet been determined (e.g. in IV, V and VI: Figure 10) would lead to trigonal methylene groups, probably methylene radicals, which exhibit torsional equivalent faces that are indistinguishable for an enzyme (Figure 13). It therefore becomes desirable to determine the steric course occurring at this second center. The substrates IV, V and VI (Figure 10) represent, however, very difficult cases. Although the methyl groups of IV and VI and 6-methylene group of V could be made chiral by isotopic substitution (a prerequisite for the determination of the steric course at these centers) the ready reversibility of the corresponding rearrangements would result in racemization of the chiral methyl groups.

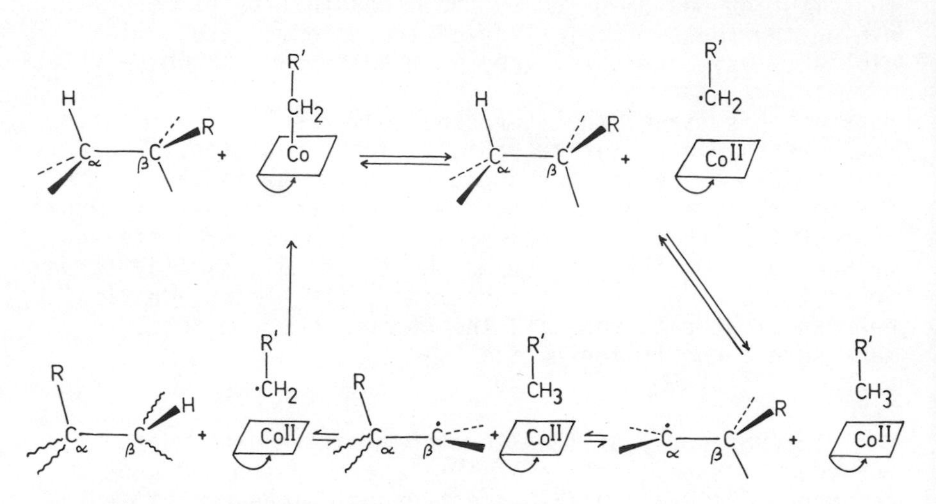

Figure 12. A working hypothesis for the mechanism of coenzyme-B_{12}-catalyzed rearrangements as elaborated on the basis of isotope labeling, electron paramagnetic resonance and electron spectroscopic studies.

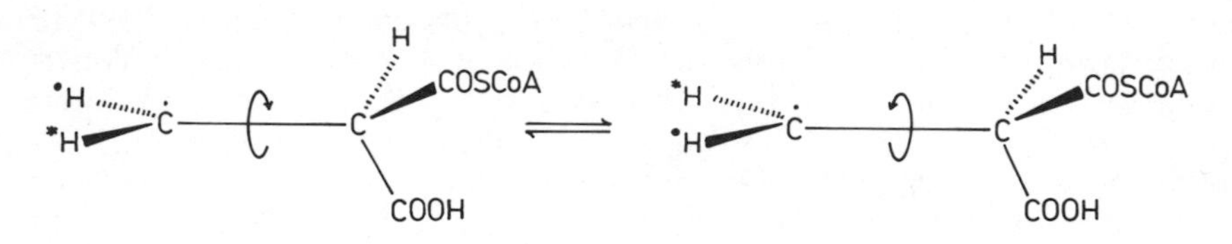

Figure 13. Stereohomotopic indistinguishable faces of the methylene radical in the hypothetical intermediate of the methylmalonyl-CoA mutase reaction.

STERIC COURSE OF THE REARRANGEMENT OF ETHYLMALONYL-CoA

The path for a solution to this problem was opened by the recent discovery[28,29] that the substrate analogue, ethylmalonyl-CoA, is also a substrate for methylmalonyl-CoA mutase from Propionibacterium shermanii. This analogue is converted to the product at only one thousandths the rate of the natural substrate, a fact which can, however, be compensated by using more enzyme and longer incubation times. In ethylmalonyl-CoA the second migration center is a prochiral methylene group and the two diastereotopic hydrogen atoms should be differentiated by the enzyme (Figure 14). The verification of this assumption required the preparation of stereospecifically labeled substrates. For the stereospecific introduction of deuterium into the ethyl group, the elegant enzymatic method of Simon and coworkers[30,31] is most suitable. This method consists of treating a primary alcohol, e.g. ethanol, in deuterium oxide with alcohol dehydrogenase and diaphorase in the presence of catalytic amounts of NAD^+. the exchange of the H_{Re}-atom in position 1 of the alcohol can be directly monitored with a high resolution NMR spectrometer (Diziol, Robinson and Retey, unpublished). In this way up to 10 g of (1R)-[1-$^2H_1$]-ethanol can be prepared at very short notice. The enantiomeric (1S)-[1-$^2H_1$]-ethanol was also obtained by exchanging [1-$^2H_2$]-ethanol in normal water. This practically useful enzymatic method is again an excellent example of chiral recognition of reflection-equivalent groups by an enzyme.

The sterospecifically deuterated ethanols were converted via their tosylates into the corresponding ethylmalonic acids. ORD examination of the resulting (3S)- and (3R)-[3-$^2H_1$]-ethylmalonic acids indicated inversion of configuration in the alkylation step. Enzymic rearrangement of the CoA esters of the two acids afforded samples of 2-methylsuccinyl-CoA from which, after hydrolysis, the acid moiety was isolated. The configuration of and the deuterium distribution in these methylsuccinic acids were determined by ORD/CD, ^1H-NMR, and mass spectrometric measurements.

Of these, the ^1H-NMR spectra were the most informative, as can be easily seen in Figure 15. At 360 MHz the tertiary hydrogen atom at C-2 and the two diastereotopic H atoms at C-3 of 2-methylsuccinate give rise to an ABX system (X is

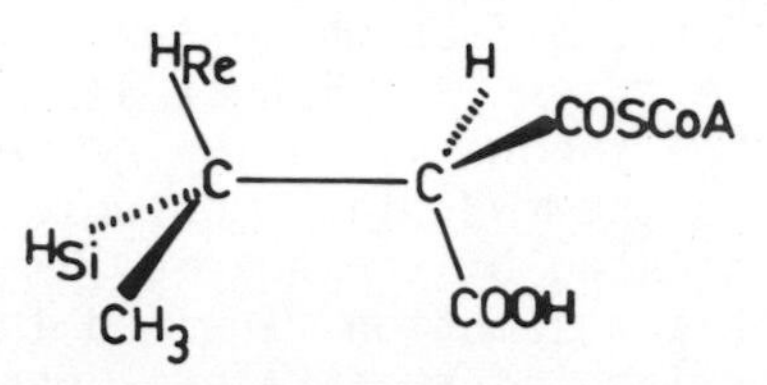

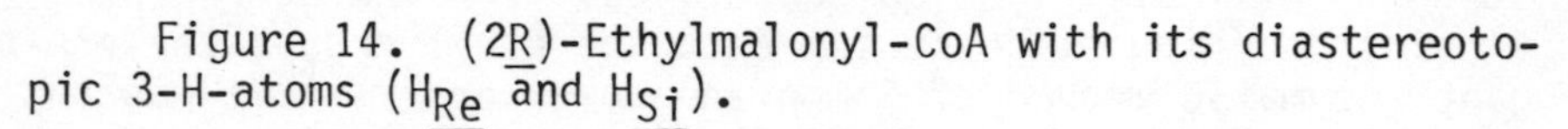

Figure 14. (2$\underline{R}$)-Ethylmalonyl-CoA with its diastereotopic 3-H-atoms ($H_{\underline{Re}}$ and $H_{\underline{Si}}$).

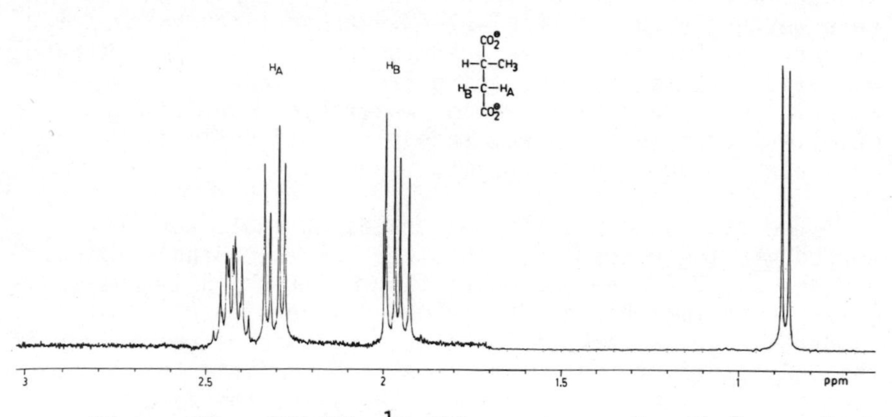

Figure 15. 360 MHz ^{1}H-NMR spectrum of methyl-succinate at pH 6 in deuterium oxide.

Table 1. Products obtained from stereospecifically deuterated ethylmalonyl-CoA species. The values are corrected for 100% deuterium content of the substrate. From (3$\underline{R}$)-[3-$^{2}H_{1}$]-ethylmalonyl-CoA 36%, unlabeled product was also obtained.

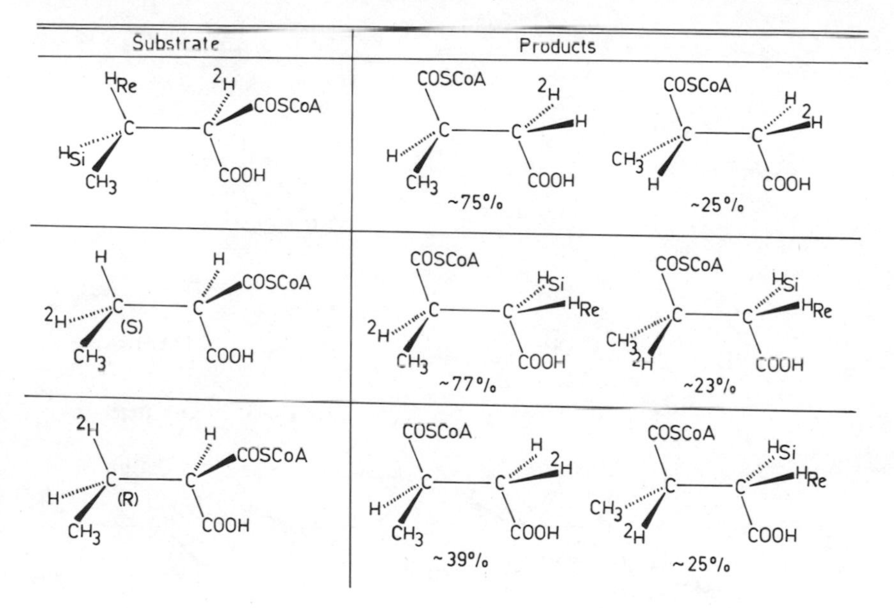

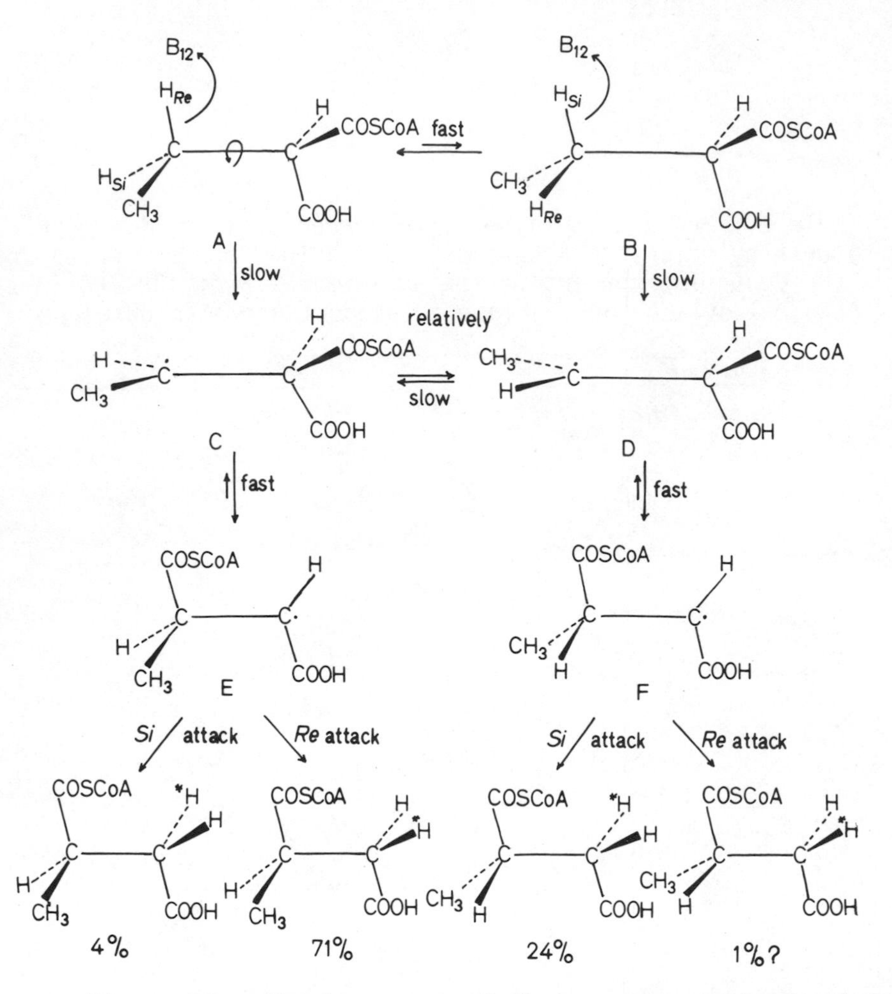

Figure 16. Steric course of the rearrangement of ethylmalonyl-CoA catalyzed by methylmalonyl-CoA mutase as elaborated by studies with stereospecifically labeled substrates. (Taken from ref. 28 with permission of Eur. J. Biochem.)

further affected by the methyl group, but this is irrelevant for our purposes). Inspection of the spectra of stereospecifically deuterated methylsuccinate specimens of known configuration allowed assignment of the 3-H_A to the quartet at lower field and of the 3-H_B to the quartet at higher field[28,29]. With this analytical tool in hand it was possible to localize the deuterium in any of the three relevant positions in the enzymatically produced methylsuccinates. A combination of ^{1}H-NMR and chiroptical measurements then allowed the results summarized in Table 1 to be drawn. The easily accessible 2-ethyl-[2-$^2H_1$]-malonyl-CoA afforded monodeuterated methylsuccinate (2R:2S ≈ 3:1) with the deuterium exclusively in the 3-erythro position (erythro with respect to the vicinal methyl group, H_A in Figure 15). From the (3S)-[3-$^2H_1$]-species, [2-$^2H_1$]-methylsuccinate (2R:2S = 77:23) was formed without loss of deuterium. Somewhat surprising results were obtained when (3R)-[3-$^2H_1$]-ethylmalonyl-CoA served as the substrate. In only about 39% of the cases did the heavy isotope migrate into the expected 3-"threo"-position and in 25% of the cases it did not migrate at all. Simultaneously 36% of the deuterium was lost during the reaction. The optical purity of this last methylsuccinic acid specimen was particularly low (2R:2S = 59:41).

We believe that these results, which unmask the infallibility of the enzyme as a stereospecific catalyst, are particularly useful in gaining a better insight into the reaction mechanism. Although there are other possible interpretations of these results, the one outlined in Figure 16 is particularly attractive and compatible with all other data. The scheme is based on the reasonable assumption that only (2R)-ethylmalonyl-CoA is accepted by the active site of methylmalonyl-CoA mutase, the configuration of which is analogous to that of the natural substrate. (The chemical synthesis of methylmalonyl-CoA and very probably also of ethylmalonyl-CoA leads to a mixture consisting of the (2R)- and (2S)-diastereoisomers in about equal amounts. Their equilibrium is readily established by heating or by a specific epimerase, also called racemase.) Accordingly, C-2 of ethylmalonyl-CoA has been assumed to be conformationally fixed at the active site (Figure 16). The site occupied by the second migration center (C-3) is originally tailored by nature, for a methyl group. Rotation about the C-C axis results in three indistinguishable geometrical situations due to the threefold torsional symmetry axis of the methyl group. In the case of ethylmalonyl-CoA these three rotamers are however geometrically and energetically different. Since the position from which hydrogen transfer to coenzyme B_{12} takes place cannot be occupied by the methyl group, two relevant conformations exist (A and B in Figure 16). The experimentally observed preferential migration of H_{Re} suggests that conformation A is favored over B. When H_{Si} is substituted by deuterium this thermodynamic effect is synergistic with a kinetic isotope effect so that practically no reaction from conformation B is observable. If, on the other hand, H_{Re} is replaced by deuterium, the predominating conformer A will react much more slowly with the result that about 25% protium will be transferred from the less favorable conformer B, leaving the deuterium in its original position.

Hydrogen abstraction from conformers A and B affords the intermediates C and D respectively (Figure 16), with a metastable trigonal carbon center (probably radical). Cobalt-catalyzed migration of the thioester group leads to the intermediate E with the (2R)-configuration when starting from conformer C, and to F with the (2S)-configuration when starting from conformer D. Back-transfer of hydrogen from the hydrogen-accepting site of the catalyst would lead to the final products. To account for the observed chiroptical data, it must further be postulated that a slow conversion of C to D takes place by rotation around the C-C bond. Thus, even in the experiment in which deuterium substitution for 3-H_{Si} largely suppressed the reaction from conformation B, about 23% of the product with (2S)-configuration was formed. Finally, the stereochemical homogeneity of the monodeuterated methylsuccinate products (the migrating hydrogen isotope ends up predominantly in the threo-position as revealed by NMR-examination) requires that the already established configuration at C-2 of intermediates E and F (Figure 14) determines which of the diastereotopic faces of the trigonal center will be attacked by the hydrogen atom. In the case of E it is mainly the Re-face, in the case of F it is mainly the Si-face.

The scheme shown in Figure 16 does not give any explanation for the substantial loss (36%) of migrating hydrogen, revealed in those experiments in which it was labeled with deuterium. The mechanistic background for this unexpected and at first confusing finding is currently under investigation in our laboratory. However, it does not belong in the scope of stereospecificity and will not be treated further here.

In summary, the stereochemical results with labeled ethylmalonyl-CoA species support the idea that in the methylmalonyl-CoA mutase reaction, intermediates with trigonal carbon centers are formed, and the stereospecificity, insofar as it has been observed, is based on the ability of the enzyme to differentiate between stereoheterotopic trigonal faces. If the enzyme is challenged by an unusual substrate, incomplete stereospecificity will result: e.g. the formation of (2*R*)- and (2*S*)-methylsuccinyl-CoA in a ratio of 3 : 1, and retention of configuration to the extent of only 70%. It is tempting to postulate that analogous trigonal intermediates will occur also in the glutamate mutase reaction. However, recognition of the stereoheterotopic faces by this enzyme will be modified in such a way that the outcome will be a net inversion of configuration[19].

STERIC COURSE OF THE AMINOETHANOL DEAMINASE REACTION

Does the existence of trigonal intermediates also hold for other coenzyme-B_{12} dependent rearrangements ? As we have already pointed out, trigonal methylene groups with torsional symmetry occurring in an enzyme-bound intermediate should exhibit indistinguishable stereohomotopic faces and should afford, upon attack by *any* reagent, racemic products. Inspection of the coenzyme-B_{12} catalyzed rearrangements reveals that such an intermediate might be formed in the reaction catalyzed by aminoethanol deaminase[35,36]. This reaction is essentially irreversible and in addition the product, acetaldehyde, can be trapped by alcohol dehydrogenase (Figure 17). Aminoethanols appropriately labeled with deuterium and tritium were prepared as outlined in Figure 18. Pyruvate-glutamate transaminase is known to exchange stereospecifically the H_{Re}-proton of glycine with solvent protons[33,34]. Incubation of [$^{2}H_{2}$]-glycine in tritiated water, and of nonspecifically tritiated glycine in deuterium oxide, with transaminase afforded the enantiomeric (2*R*)- and (2*S*)-[^{2}H, ^{3}H]-glycines. The tritium was present at a tracer level only, but it was ensured that all molecules containing tritium also contained deuterium. The doubly labeled glycines were transformed into the corresponding aminoethanols,

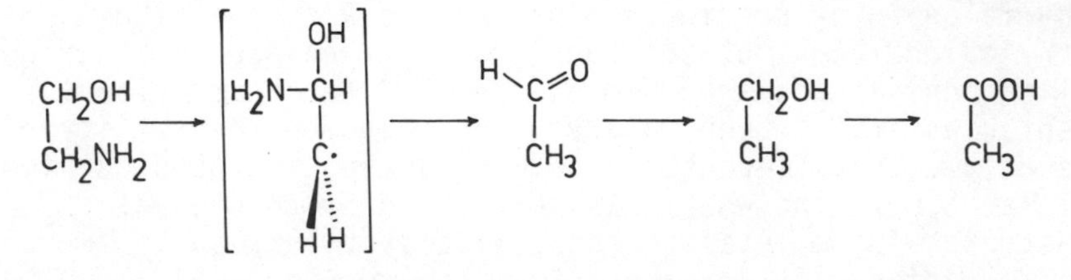

Figure 17. Conversion of aminoethanol to acetaldehyde by the coenzyme-B_{12}-dependent deaminase, subsequent trapping of the product with alcoholdehydrogenase, and oxidation of the ethanol to acetic acid.

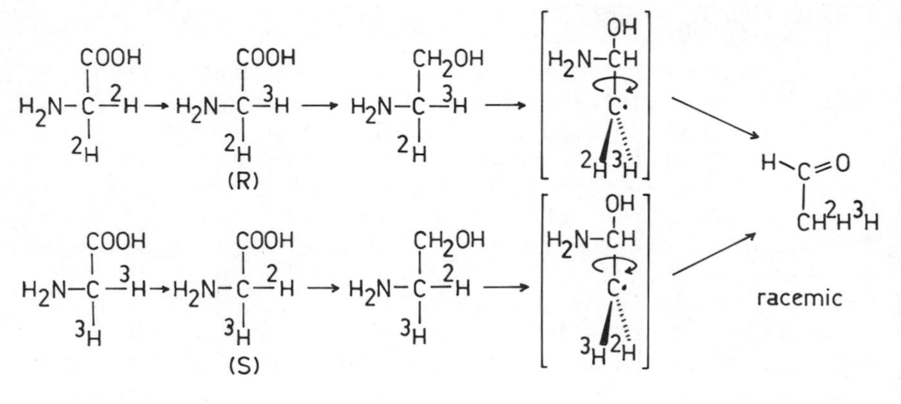

Figure 18. Preparation of (2$\underline{R}$)- and (2$\underline{S}$)-[2-$^{2}H_{1}$,$^{3}H_{1}$]-aminoethanols and their conversion to racemic [2-$^{2}H_{1}$,$^{3}H_{1}$]-acetaldehyde on aminoethanol deaminase.

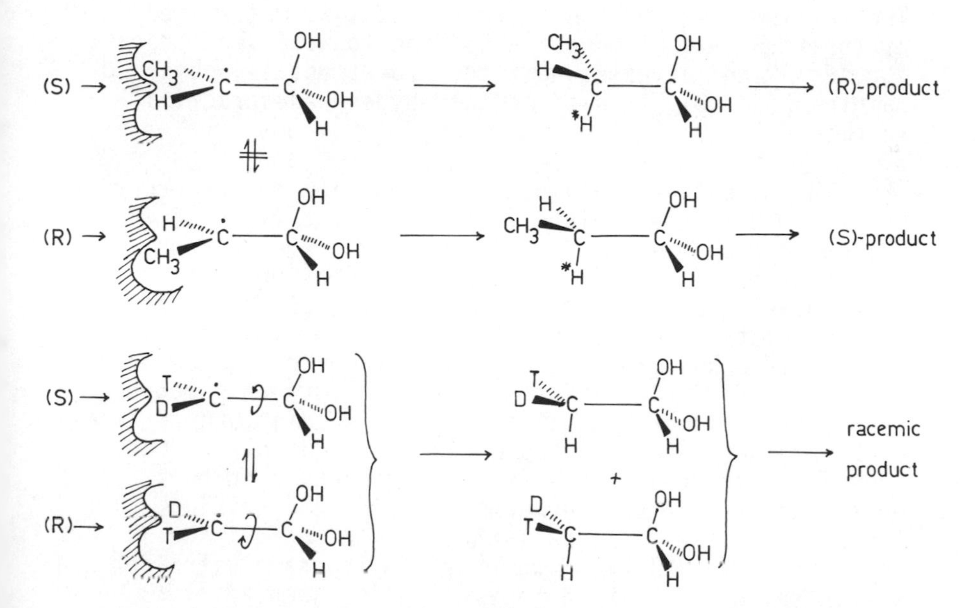

Figure 19. The steric course of the dioldehydrase reaction with (2$\underline{S}$)- and (2$\underline{R}$)-propane-1,2-diols and with (S)- and ($\underline{R}$)-[$^2H_1$ $^3H_1$]-ethylene glycols.

which then served as substrates for the coenzyme-B_{12} dependent deaminase from an unclassified *Clostridium*[35,36]. Stereospecific substitution of the amino group by protium would lead to the formation of stereochemically pure chiral methyl groups.

The chiral methyl groups were analyzed by taking the produced [$^2H_1$, $^3H_1$]-acetaldehyde through the sequence: acetaldehyde ⟶ ethanol ⟶ acetic acid ⟶ acetyl CoA ⟶ malate using the appropriate enzymes or reagents. The tritium distribution in the malate was determined by treatment with fumarase according to published procedures[37,38] and revealed that the experimentally obtained samples of doubly labeled acetaldehydes were practically racemic.[39]

This result can again be rationalized by the occurrence of a trigonal intermediate (Figure 18) the faces of which have been made distinguishable for a subsequent sophisticated enzymatic analysis by isotopic labeling, but which were indistinguishable for the aminoethanol deaminase. Such a constellation is possible because the enzymatic analysis of chiral methyl groups is based on *kinetic* isotope effects being significantly different for the hydrogen isotopes, whereas the recognition of the formally enantiotopic faces of a trigonal C-^{2}H,^{3}H group would be based only on *thermodynamic* isotope effects (e.g. differences in bulk, polarizability, etc.) between deuterium and tritium. The latter effects are so small that such a recognition is beyond the capability of an enzyme. The same situation arises in the case of the coenzyme-B_{12} dependent diol dehydrase. This enzyme converts (*R*)- and (*S*)-propane-1,2-diol stereospecifically with inversion[16,17] to propionaldehyde. Here the stereospecificity is probably due to a hindered rotation of the enzyme-bound trigonal intermediate, Figure 19. The shorter substrate analogue, ethylene glycol, affords acetaldehyde with racemization[18], as shown by appropriate labeling with hydrogen isotopes. The latter results demonstrate in a particularly impressive manner, that a schematic interpretation of the steric course of enzymatic reactions in mechanistic terms is useless, since the same enzyme converts two structurally related substrates such as propane-1,2-diol and ethylene glycol with different stereochemical courses.

CONCLUSIONS

In the past 15 years a considerable body of experimental material on enzymatic stereospecificity has been collected, a complete review of which is beyond the scope of this chapter. The benefit of these data for our understanding of the corresponding reaction mechanisms was disproportionately small compared with the extremely delicate work involved and the tremendous amount of effort invested. In the beginning, admiration of and homage to a high degree of stereospecificity prevailed. But now we are beginning to understand that the steric course of most enzymatic processes, normally the best indicator for the reaction mechanism, is obscured by chiral recognition. Disentangling these two factors is difficult, but not impossible. The most useful information as to the mechanism can be obtained when an enzyme is "fooled" in such a way, for example by appropriate isotopic labeling or the use of substrate analogues that the metastable intermediate exhibits rotation-equivalent (stereohomotopic) faces which are indistinguishable even for the enzyme. A better insight into chiral recognition by enzymes, which is based mainly on very intimate nonbonding or ionic interactions, has to wait until the exact geometry of the relevant active sites becomes known.

ACKNOWLEDGEMENTS

I thank the Deutsche Forschungsgemeinschaft, the Fonds der Chemischen Industrie, and the Schweizerische Nationalfonds for generous financial support. This chapter could not have been prepared without the enthusiastic help and collaboration of my colleagues mentioned in the References. I am particularly grateful to Dr. J. A. Robinson for improving the English style of the manuscript.

REFERENCES

1. Hanson, K. R. and I. A. Rose 1963. The absolute stereochemical course of citric acid biosynthesis. Proc. Natl. Acad. Sci. U.S. 50:981-988.
2. Weber, H. 1965. Untersuchungen zum sterischen Verlauf enzymatischer Reaktionen an Substraten mit Meso-Kohlen stoffatom. Doctoral Thesis, Eidgenossische Technische Hochschule Zurich No. 3591.
3. Retey, J., K. Bartl, E. Ripp and W. E. Hull. 1977. Stereospecificity of phenylpyruvate tautomerase. Eur. J. Biochem. 72:251-257.
4. Bartl, K., C. Cavalar, T. Krebs, E. Ripp, J. Retey, W. E. Hull, H. Gunther and H. Simon. 1977. Synthesis of stereospecifically deuterated phenylalanines and determination of their configuration. Eur. J. Biochem. 72:247-250.
5. Poston, J. M. 1978. Coenzyme-B_{12}-dependent enzymes in potatoes: leucine 2,3-aminomutase and methylmalonyl-CoA mutase. Phytochemistry 17:401-402.
6. Kellermeyer, R. W. and H. G. Wood. 1962. Methylmalonyl isomerase: a study of the mechanism of isomerization. Biochemistry 1:1124-1131.
7. Phares, E. F., M. V. Long and S. F. Carson. 1962. An intramolecular rearrangement in the methylmalonyl isomerase reaction as demonstrated by positive and negative ion mass analysis of succinic acid. Biochem. Biophys. Res. Commun. 8:142-146.
8. Cardinale, G. J. and R. H. Abeles. 1967. Mechanistic similarities in the reactions catalyzed by dioldehydrase and methylmalonyl-CoA mutase. Biochim. Biophys. Acta 132:517-518.
9. Frey, P. A. and R. H. Abeles. 1966. The role of the B_{12} coenzyme in the conversion of 1,2-propanediol to propionaldehyde. J. Biol. Chem. 241:2732-2833.
10. Retey, J. and D. Arigoni. 1966. Coenzym-B_{12} als gemeinsamer Wasserstoffubertrager der Dioldehydrase- und der Methylmalonyl-CoA-Mutase Reaktion. Experientia 22:783-784.
11. Abeles, R. H. and D. Dolphin. 1976. The vitamin B_{12} coenzyme. Acc. Chem. Res. 9:114-120.
12. Barker, H. A. 1972. Corrinoid-dependent enzymic reactions. Annu. Rev. Biochem. 41:55-90.

13. Hogenkamp, H. P. C. 1968. Enzymatic reactions involving corrinoids. Annu. Rev. Biochem. 37:225-248.
14. Schrauzer, G. N. 1977. Neuere Entwicklungen auf dem Gebiet des Vitamins B_{12}: Von einfachen Corrinen and von coenzym-B_{12}-abhangigen Enzymreaktionen. Angew. Chem. 77:239-251; Angew. Chem. Int. Ed. 16:233-245.
15. Stadtman, T. C. 1971. Vitamin B_{12}. Science 171:859.
16. Retey, J., A. Umani-Ronchi, and D. Arigoni. 1966. Zur Stereochemie der Propandioldehydrase-Reaktion. Experientia 22:72-73.
17. Zagalak, B., P. A. Frey, G. L. Karabatsos, and R. H. Abeles. 1966. The stereochemistry of the conversion of D and L 1,2-propane-diols to propionaldehyde. J. Biol. Chem. 241:3028-3055.
18. Bonetti, V. 1974. Etude Stereochimique des reactions catalysees par les deshydrases utilisant la coenzyme B_{12}. Doctoral Thesis, Eidgenossische Technische Hochschule Zurich No. 5366.
19. Sprecher, M., R. L. Switzer and D. B. Sprinson. 1966. Stereochemistry of the glutamate mutase reaction. J. Biol. Chem. 241:864-867.
20. Retey, J., F. Kunz, D. Arigoni, T. C. Stadtman. 1978. Zur Kenntnis der β-Lysin-Mutase-Reaktion: Mechanismus und sterischer Verlauf. Helv. Chim. Acta 61:2989-2998.
21. Sprecher, M., M. S. Clark, and D. B. Sprinson. 1966. The absolute configuration of methylmalonyl coenzyme A and stereochemistry of the methylmalonyl coenzyme A mutase reaction. J. Biol. Chem. 241:872-877.
22. Retey, J., A. Umani Ronchi, J. Seibl, and D. Arigoni. 1966. Zum Mechanismus der Propanidoldehydrase-Reaktion. Experientia 22:502-503.
23. Babior, B. M., T. H. Moss, and D. C. Gould. 1972. The mechanism of action of ethanolamine ammonia lyase, a B_{12}-dependent enzyme. J. Biol. Chem. 247:4389-4392.
24. Babior, B. M., T. H. Moss, W. H. Orme-Johnson, and H. Beinert. 1974. The mechanism of action of ethanolamine ammonia-lyase, a B_{12}-dependent enzyme. J. Biol. Chem. 249:4537-4544.
25. Cockle, S. A., H. A. O. Hill, R. J. P. Williams, D. P. Davies, and M. A. Foster. 1972. The detection of intermediates during the conversion of

propane-1,2-diol to propionaldehyde by glycerol dehydrase, a coenzyme-B_{12}- dependent enzyme. J. Am. Chem. Soc. 94:275-277.

26. Finlay, T. H., J. Valinsky, A. S. Mildvan, and R. H. Abeles. 1973. Electron spin resonance studies with diol dehydrase. J. Biol. Chem. 248:1285-1290.
27. Joblin, K. N., A. W. Johnson, M. F. Lappert, M. R. Hollaway, and H. A. White. 1975. Coenzyme-B_{12}-dependent enzyme reactions: a spectrophotometric rapid kinetic study of ethanolamine ammonia lyase. FEBS Lett. 53:193-198.
28. Retey, J., E. H. Smith, and B. Zagalak. 1978. Investigation of the mechanism of the methylmalonyl-CoA mutase reaction with the substrate analogue: Ethylmalonyl-CoA. Eur. J. Biochem. 83:437-451.
29. Retey, J. and B. Zagalak. 1973. Stereochemie der coenzym-B_{12}-abhangigen Methylmalonyl-CoA-Mutase-Reaktion. Untersuchungen mit Aethylmalonyl-CoA. Angew. Chem. 85:721-722; Angew. Chem. Int. Ed. 12:671-672.
30. Gunther, H., M. A. Alizade, M. Kellner, F. Biller and H. Simon. 1973. Preparation of (1R)[1-^{2}H]- and (1S)[1-^{2}H]-alcohols by exchange reactions catalyzed by yeast or a coupled enzyme system. Z. Naturforsch. 28c:241-246.
31. Gunther, H., F. Biller, M. Kellner and H. Simon. 1973. Praparative Darstellung von (1R)- und (1S)-Monodeuteriopropanol durch enzymatische Austauschreaktionen. Angew. Chem. 85:141-142; Angew. Chem. Int. Ed. 12:146-147.
32. Arigoni, D. and E. L. Eliel. 1969. Chirality due to the presence of hydrogen isotopes at non-cyclic positions. Top. Stereochem. 4:192-199.
33. Akthar, M. and P. M. Jordan. 1969. The absolute configuration of stereospecifically tritiated glycines. Tetrahedron Lett. 875-877.
34. Besmer, P. and D. Arigoni. 1968. Stereochemische Untersuchungen mit chiral markiertem Glycin. Chimia 22:494.
35. Babior, B. M. 1969. The mechanism of action of ethanolamine deaminase. J. Biol. Chem. 244:449-456.
36. Kaplan, B. H. and E. R. Stadtman. 1968. Ethanolamine deaminase, a cobamide coenzyme-dependent enzyme. J. Biol. Chem. 243:1794-1803.
37. Cornforth, J. W., J. W. Redmond, H. Eggerer, W. Buckel, and Ch. Gutschow. 1969. Asymmetric methyl

groups and the mechanism of malate synthase. Nature 221:1212-1213.

38. Luthy, J., J. Retey, and D. Arigoni. 1969. Preparation and detection of chiral methyl groups. Nature 221:1213-1215.

39. Retey, J., C. J. Suckling, D. Arigoni, and B. M. Babior. 1974. The stereochemistry of the reaction catalyzed by ethanolamine ammonia-lyase, an adenosylcobalamin-dependent enzyme. J. Biol. Chem. 249:6359-6360.

Chapter Two

STEREOCHEMICAL STUDIES ON THE METABOLISM OF AMINO ACIDS

D. J. ABERHART

The Worcester Foundation for Experimental Biology
Shrewsbury, Massachusetts

INTRODUCTION

This chapter is devoted to the discussion of two topics concerning the stereochemistry of the metabolism of branched-chain amino acids. First we will review our work on the stereochemistry of the biosynthesis of penicillins from valine in *Penicillium chrysogenum*. The second topic will be the stereochemistry of certain steps in the degradative metabolism of L-valine and L-leucine in several organisms.

STEREOCHEMISTRY OF THE BIOSYNTHESIS OF PENICILLINS

Figure 1 summarizes the so-called tripeptide theory[1] of the biosynthesis of pencillins, exemplified by the formation of penicillin N. A tripeptide, **1**, consisting of L-α-amino-adipic acid, L-cysteine, and D-valine, abbreviated ACV,

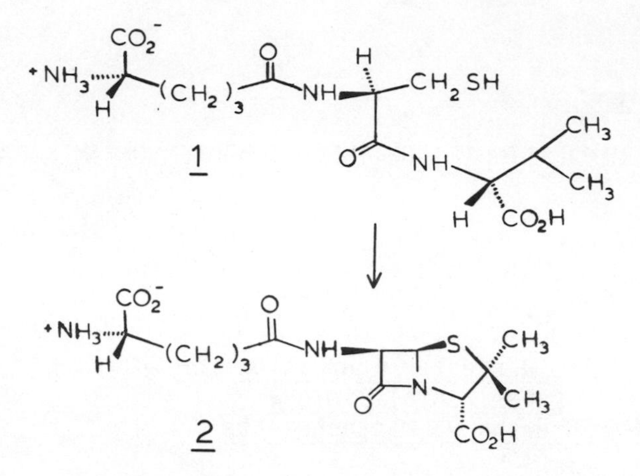

Figure 1. The tripeptide theory illustrating the biosynthesis of penicillin N from L-α-aminoadipyl-L-cysteinyl--D-valine (LLD-ACV) in Penicillium chrysogenum.

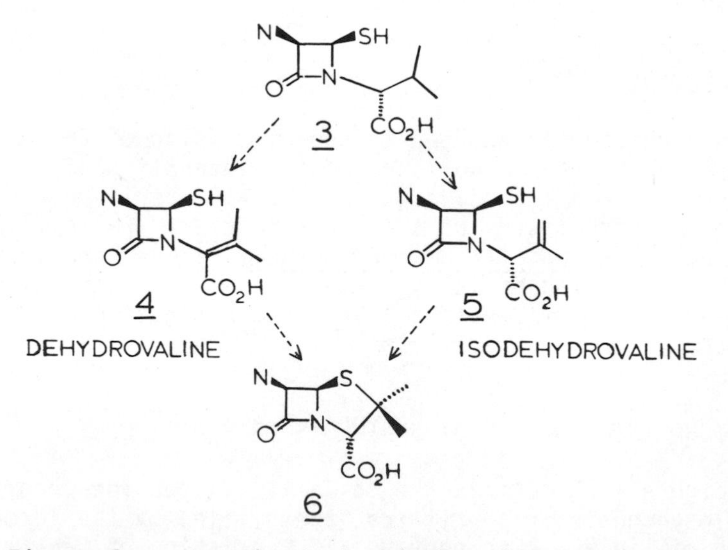

Figure 2. Hypothetical dehydrovaline or isodehydrovaline intermediates in the biosynthesis of penicillins.

undergoes cyclization to the penam ring system 2 by an unknown mechanism. This tripeptide has been isolated from *Penicillium chrysogenum*[1] and *Cephalosporium* spp[2] (which produce penicillin N and the related compound cephalosporin C). The ACV tripeptide is convertible by a cell-free system, obtained by osmotic lysis of protoplasts from *C. acremonium*, into penicillin N[3]. Although the tripeptide contains D-valine, it is apparently first formed from L-valine and not D-valine. The inversion of configuration occurs *without* loss of the original nitrogen atom of L-valine.

The lactam ring. Obviously in the conversion of the ACV tripeptide to the penam ring system, bond formation must occur between the D-valine nitrogen and the β-carbon of cysteine to form the β-lactam ring, and between the cysteine sulfhydryl group and the β-carbon of valine to form the thiazolidine ring. It has generally been assumed that the β-lactam ring is the first to be formed, but in fact there is no biochemical evidence to support this hypothesis.

The thiazolidine ring. When our work was initiated, the prevailing theory for the formation of the thiazolidine ring of penicillin was that the valine unit of a β-lactam-containing intermediate (3) underwent dehydrogenation to form an α,β-dehydrovaline intermediate, 4 (Figure 2). However, such dehydrovaline intermediates had never been detected in systems producing penicillins. Synthetic α,β-dehydrovaline intermediates of the type shown in Figure 2 failed to undergo cyclization to the penam ring system *in vitro*, probably as the result of an unfavorable alignment of pertinent orbitals[4]. Also, it has recently been shown that conversion of ACV having an α-^{3}H atom in the valine residue to penicillin N in the cell-free system mentioned above proceeds without loss of tritium[3]. Therefore, it seems probable that (at least) non-enzyme-bound α,β-dehydrovaline intermediates do not participate in penicillin biosynthesis.

As an alternative, we considered the possible involvement of the β,γ-dehydrovaline intermediate (isodehydrovaline, 5) shown in Figure 2. Sulfenic acid analogues of this structure have been cyclized chemically to penicillins and/or cephalosporins[5]. To test the possible involvement of isodehydrovaline derivatives in penicillin biosynthesis, we synthesized [methyl-$^{2}H_{6}$] valine, 7 (Figure 3). This was converted by *P. chrysogenum* into phenoxymethylpenicillin

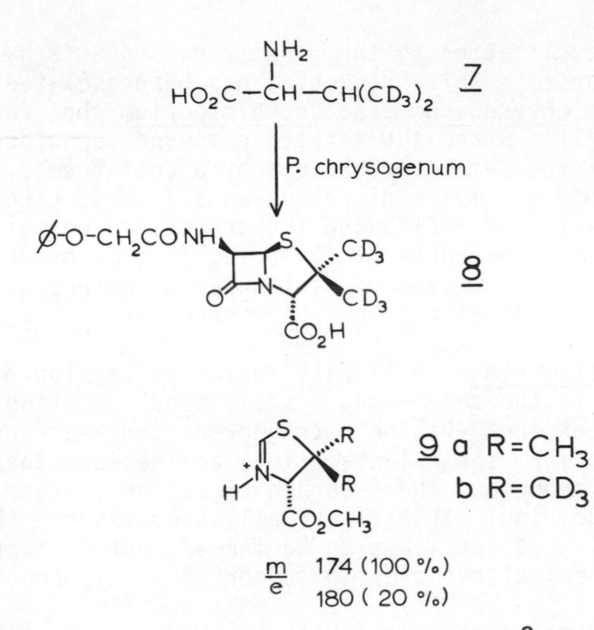

Figure 3. Incorporation of DL-[methyl$^{2}H_6$] valine into penicillin V. Mass spectrometric evidence indicating the retention of all six deuterons in the biosynthetic penicillin.

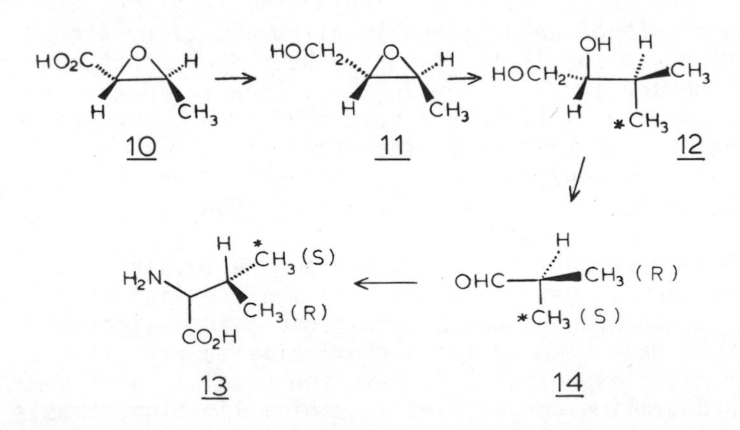

Figure 4. Synthetic pathway used for the synthesis of (2RS,3S) chirally labeled valines.

(penicillin V), 8. The biosynthetically produced penicillin V was converted with diazomethane into its methyl ester and analyzed by mass spectrometry. The spectrum of the unlabeled penicillin V methyl ester has a very intense base peak at m/e 174, 9a. In the penicillin V methyl ester synthesized from [$^{2}H_{6}$] valine the m/e 174 peak was accompanied by a new peak at m/e 180 (20%), 9b (Figure 3). It can be concluded, therefore, that penicillin biosynthesis proceeds without exchange of any of the valine methyl hydrogens, thus ruling out non-enzyme bound isodehydrovaline intermediates[6].

We then turned our attention to the stereochemistry of the cyclization reaction which lead to the formation of the thiazolidine ring of penicillin. We assumed a priori that bond formation between the cysteine sulfhydryl group and the β-carbon of valine would proceed stereospecifically with either retention or inversion of configuration. For the evaluation of the stereochemistry of the reaction, we synthesized ^{13}C-chirally-labeled valine ("chiral valine") and planned to incubate this with P. chrysogenum and isolate the biosynthesized penicillin. The position of ^{13}C enhancement in the penicillin could be determined by ^{13}C NMR, since the upfield and downfield ^{13}C methyl signals of penicillins had earlier been assigned to the α and β-methyl carbons, respectively[7].

The required chiral valine was synthesized as outlined in Figure 4[8,9]. This synthesis was first modeled using deuterium labeling and then applied to the ^{13}C analogue. At about the same time, syntheses of chiral valines were reported independently by Hill[10], Baldwin[11,12], and Sih[13] and their coworkers. In our synthetic route, the optically pure epoxy acid 10 was converted to the methyl ester, which was then reduced with sodium borohydride to the sensitive epoxy alcohol 11 in moderate yield. This alcohol was immediately treated with labeled methyllithium to yield the glycol, 12. The conditions for the opening of the oxirane ring are critical, and most importantly the labeled methyllithium must be halide-free. The labeled methyllithium was prepared by halide exchange of labeled methyl iodide with n-butyllithium in hexane. The attempted use of methyllithium prepared from methyl iodide and lithium metal was unsuccessful, since the lithium iodide present caused epoxide rearrangements to occur faster than the opening of the oxirane ring. Even with halide-free methyllithium, the reaction was not clean,

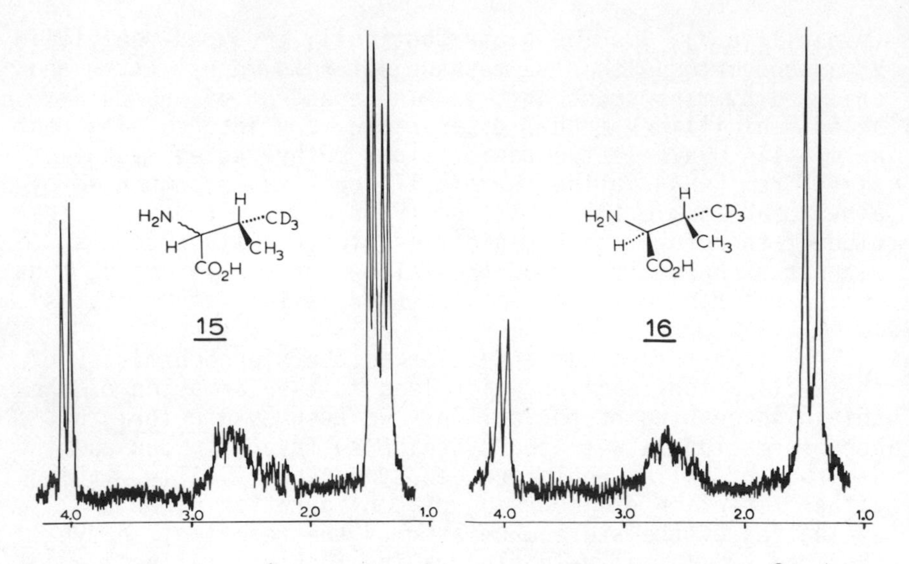

Figure 5. ^1H NMR spectra of (2RS,3S)-[4,4,4-$^2H_3$] valine and (2S,3S)-[4,4,4-$^2H_3$] valine (^2H_2O solutions).

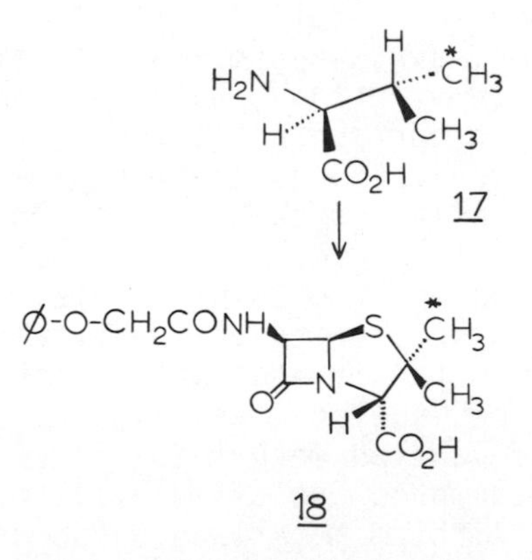

Figure 6. Summary of the stereochemistry of the incorporation of chirally labeled valine into penicillin V.

and the product difficult to purify. Fortunately, however, the impurities do not interfere with the subsequent steps. Cleavage of the glycol 12 with periodate gives chiral isobutyraldehyde, 13, (not isolated), which was subjected to a Strecker synthesis to yield DL-chiral valine, 14. This method has been used for the synthesis of (S)-[CD_3], (S)-[$^{13}CH_3$]- and both (R) and (S)-[C^3H_3]-chiral valines[8,9]. As will be seen later, the sequence has also been useful for the synthesis of chiral isobutyrates.

To prove the configurational purity of deuterated chiral valine the product was acetylated and treated with acylase, yielding L-valine and D-valine acetate. The NMR of the L-valine [$^2H_3$], 16, Figure 5, showed only one methyl doublet, whereas before resolution, 15, two methyl doublets of half this intensity were present. Similarly only one methyl doublet was seen in the $Eu(dpm)_3$ shifted NMR of the D-valine acetate (after methylation with diazomethane). The methyl signals of unlabeled valine acetate methyl ester become completely resolved in the presence of $Eu(dpm)_3$.

The (3S)-[^{13}C]-chiral valine was incubated with *P. chrysogenum* to yield penicillin V. The ^{13}C NMR of the product showed a 2.5-fold enhancement of the upfield methyl signal assigned to the α-methyl group, while the remainder of the spectrum was unchanged. The results show that bond formation between the sulfur of cysteine and the β-carbon of valine occurs with retention of configuration, as summarized in Figure 6, (17 ⟶ 18). Similar findings were independently reported by Baldwin *et al.*[12] using (3R)-[^{13}C]-chiral valine, and by Sih *et al.*[13] for the biosynthesis of penicillin N in *C. acremonium*.

The mechanism of the elaboration of the thiazolidine ring of penicillins remains unknown. One possibility to be considered is the displacement of a suitable leaving group, X, located at the β-carbon of valine, by a sulfide anion, as shown in Figure 7 (19, 20, 21). Such a process would be expected on mechanistic grounds to proceed with inversion of configuration. Thus, to be compatible with the stereochemical results discussed above, it would be necessary to demonstrate that such a leaving group was also *introduced* with inversion of configuration. In this connection it may be mentioned that Abraham and Loder[2] have isolated from *Cephalosporium sp.* a tetrapeptide consisting of α-aminoadipic

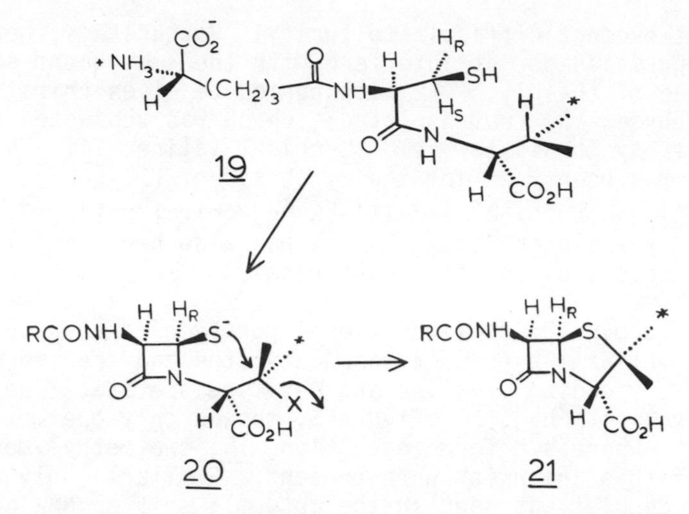

Figure 7. Hypothetical mechanism for thiazolidine ring closure in penicillin biosynthesis, illustrating overall retention of configuration at C-3 of both cysteine and valine precursors.

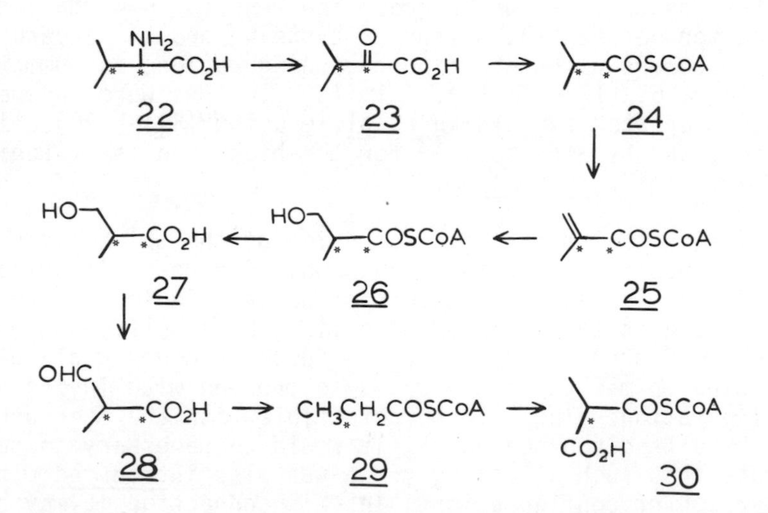

Figure 8. Pathway for degradation (catabolism) of L-valine.

acid, cysteine, β-hydroxyvaline, and glycine. Whether or not the β-hydroxyvaline in this tetrapeptide is involved in the biosynthesis of penicillin remains to be determined.

It should also be noted that we[14] and Young *et al.*[15-17] have shown that the formation of the β-lactam ring in penicillins occurs with retention of configuration at the β-carbon of the cysteine unit as shown in Figure 7 (19 ⟶ 20). Thus, both ring closures in the biosynthesis of penicillin proceed with retention of configuration.

STEREOCHEMICAL STUDIES IN THE CATABOLISM OF BRANCHED CHAIN AMINO ACIDS

Valine catabolism. The catabolic pathway for L-valine that apparently operates in all organisms is summarized in Figure 8, (22 ⟶ 30). A variant of this pathway was proposed some years ago by Coon *et al.*[18], who suggested that the intermediate methylmalonate semialdehyde, 28, underwent direct oxidation to methylmalonate, 30, without the obligatory intermediacy of propionyl CoA, 29. However, Tanaka *et al.*[19-21] have shown that [2,3-$^{13}C_2$] valine was converted by a human patient with methylmalonic acidemia to methylmalonate with retention of only one atom of ^{13}C, that from C-3 of the valine, as shown in Figure 8 (the position of ^{13}C labeling in Tanaka's experiment is shown by an asterisk). The results indicate that, at least in this case, the intermediate methylmalonyl semialdehyde must be decarboxylated before conversion to methylmalonate. Other published evidence strongly supports the operation of the same pathway in bacteria[22].

We were initially interested in determining which of the diastereotopic methyls of valine underwent oxidation during the degradation of the amino acid. For this purpose, the (3*R*) and (3*S*) chiral valines, 31 and 32 respectively, (Figure 9) labeled with tritium in one of the methyl groups were synthesized by the route previously described. The catabolic pathway outlined in Figure 8 indicates that both methyl carbons of valine are retained up to the stage of methylmalonate. However, the hydrogen atoms of one of the methyls are completely lost by its oxidation to a carboxyl group. Thus, we expected that upon conversion to propionate or subsequent metabolites, one of the tritiated methyl groups of the chiral valines would show a total loss of tri-

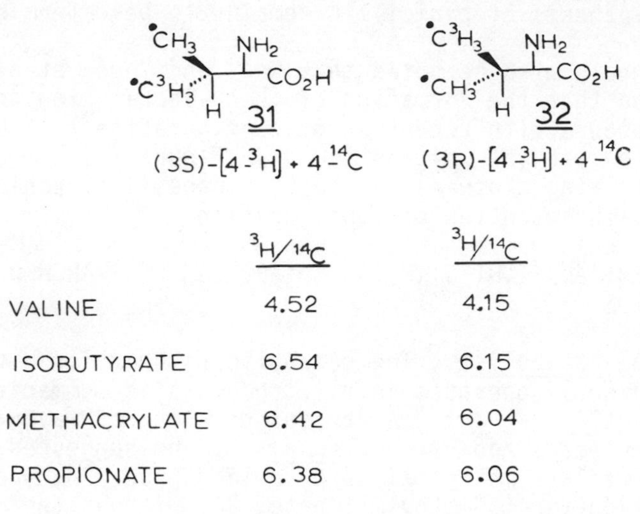

	$^3H/^{14}C$	$^3H/^{14}C$
VALINE	4.52	4.15
ISOBUTYRATE	6.54	6.15
METHACRYLATE	6.42	6.04
PROPIONATE	6.38	6.06

Figure 9. $^3H/^{14}C$ Ratios of metabolites isolated after incubation of (3R) or (3S)-[4-^3H]-+[4,4'-^{14}C]-valine with a rat liver homogenate.

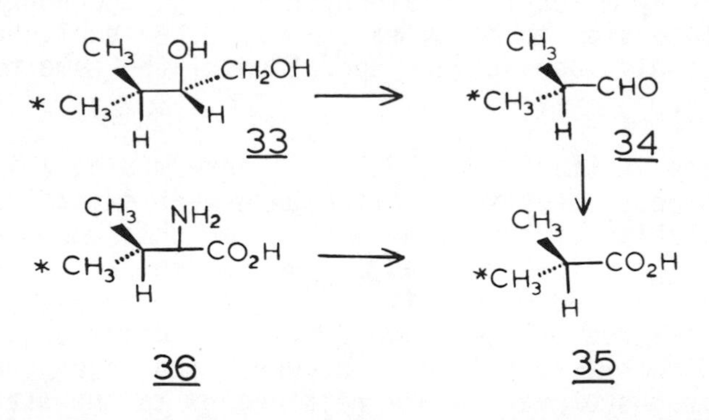

Figure 10. Pathway used for the synthesis of (2S) chirally labeled isobutyric acids.

tium while in the other tritium would be completely retained. In order to determine this the two tritiated chiral valines (31 and 32, Figure 9) were mixed with [4,4'-^{14}C]-valine, and incubated with a rat liver homogenate. Subsequently the metabolites, isobutyrate, methacrylate, and propionate were isolated, and the ^{3}H/^{14}C ratios were determined. The results are shown in Figure 9.

The metabolites from both precursors showed an increase in ^{3}H/^{14}C ratio. Presumably some hydrogen isotope effect is operating to cause the accumulation of tritium-labeled catabolic intermediates. Surprisingly, the propionate from neither (3*R*) nor (3*S*)-[^{3}H]-valine showed a loss of tritium. The ^{3}H/^{14}C ratios of all three isolated metabolites were essentially identical, and thus the oxidation was apparently nonselective with regard to the two methyl groups of valine. In a similar experiment, Tanaka *et al.*[23] have administered (3*S*)-[^{13}C]-chiral valine to vitamin B_{12} and folate-deficient rats, and examined the methylmalonate isolated from the urine by mass spectrometry. The ^{13}C label in this metabolite was distributed equally between the methyl group and one of the carboxyl groups. The [^{13}C]-valine used in this experiment was from the same preparation used in our penicillin studies. Thus, the *apparent* lack of stereospecificity of the oxidation of valine to methylmalonate cannot be attributed to possible nonstereospecific labeling of the precursor. The most reasonable explanation of the results of both experiments is that the intermediate α-ketoisovalerate[23] undergoes enolization at a rate exceeding that of its decarboxylation to isobutyryl CoA, (24) and thereby the configurational purity of the labeled precursor is lost.

Isobutyrate catabolism. In order to bypass this configurationally labile intermediate, we therefore investigated the stereochemistry of methyl group oxidation in this pathway using chirally labeled isobutyrate. The synthesis of chiral isobutyrates is outlined in Figure 10. The starting material, 33, is the intermediate used in the synthesis of the chiral valines (see Figure 4). Two oxidative steps (periodate cleavage to chiral isobutyraldehyde, 34, followed by treatment with neutral aqueous permanganate) gave the chiral isobutyric acid 35 in good yield[24]. This product could also be obtained in one step from chiral valine 36 by oxidation with silver(II) oxide[25].

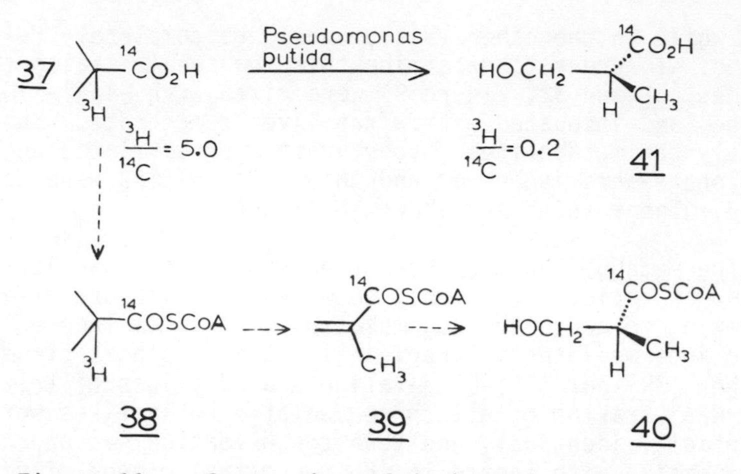

Figure 11. Conversion of isobutyric acid to β-hydroxyisobutyric acid (β-HIBA) by Pseudomonas putida (ATCC 21244). Evidence establishing the involvement of an α,β-dehydro intermediate in the pathway.

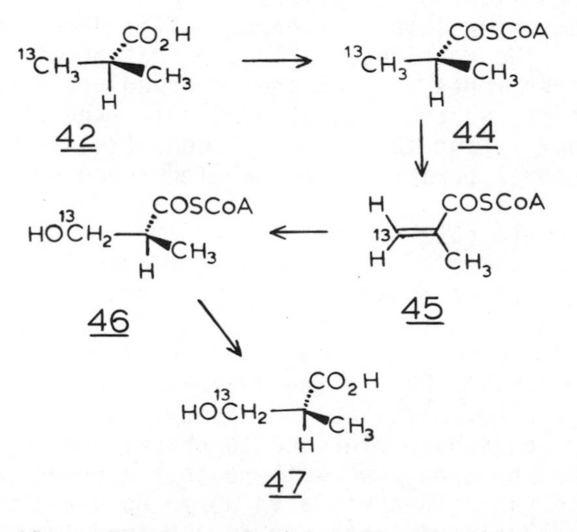

Figure 12. Summary of the stereochemistry of the conversion of chiral isobutyrate to β-HIBA by Ps. putida: oxidation of the 2-pro-S methyl group.

Since only very small quantities of the products of the catabolism of valine or isobutyrate accumulate in rats or rat liver homogenates, we considered other organisms for studying the stereochemistry of the metabolism of chiral isobutyrate. In this connection, we were particularly attracted to a report by a group at Eastman Kodak Laboratories[26] that a strain of Pseudomonas putida, grown on isobutyrate as sole carbon source, accumulated (S)(+)-β-hydroxyisobutyric acid (β-HIBA) in the medium in yields of up to 40%. Since β-HIBA is a normal intermediate of valine catabolism, this organism appeared useful for our purposes. It was established[27] that α-tritiated isobutyrate (37) was converted to β-HIBA (41) with loss of 96% of the tritium, as shown in Figure 11. This is consistent with the view that the metabolism proceeded via an α, β-dehydro intermediate (39) rather than by direct hydroxylation of one of the methyl groups.

Thus, (2S)-[3-^{13}C]-isobutyrate (42) was incubated with washed cells of Ps. putida, and the resultant β-HIBA (47) isolated as the benzoate of the methyl ester (43 Figures 12 and 13). The ^{13}C NMR spectra of the biosynthetic product (42) and an unlabeled sample, 43, clearly show the enhancement of ^{13}CX signal intensity only in the hydroxymethyl group (D in Figure 13). The relative intensities of the other signals are unchanged. The results summarized in Figure 12, (42 ⟶ 47) show that the conversion of isobutyrate to β-HIBA by this organism proceeds stereospecifically, via an unsaturated intermediate (probably methacrylyl CoA, 45), and that it is the 2-pro-S methyl group of isobutyrate which is oxidized. Also the results show that the hydration of the intermediate methacrylyl CoA proceeds stereospecifically. The addition of the hydrogen atom at C-2 takes place on the re face of the double bond, which is the same face from which it is removed in the dehydrogenation step[27]. A similar experiment in rat liver using chiral [^{3}H]-isobutyrates, prepared by silver(II) oxide oxidation of chiral [^{3}H]-valines is currently in progress.

Since the above results showed that the hydration of methacrylyl CoA in Ps. putida proceeded stereospecifically at C-2, we became interested in determining the stereochemistry of hydroxyl addition at C-3. For this purpose, we planned to synthesize stereospecifically deuterated methacrylic acid, convert it to β-HIBA by Ps. putida, and deter-

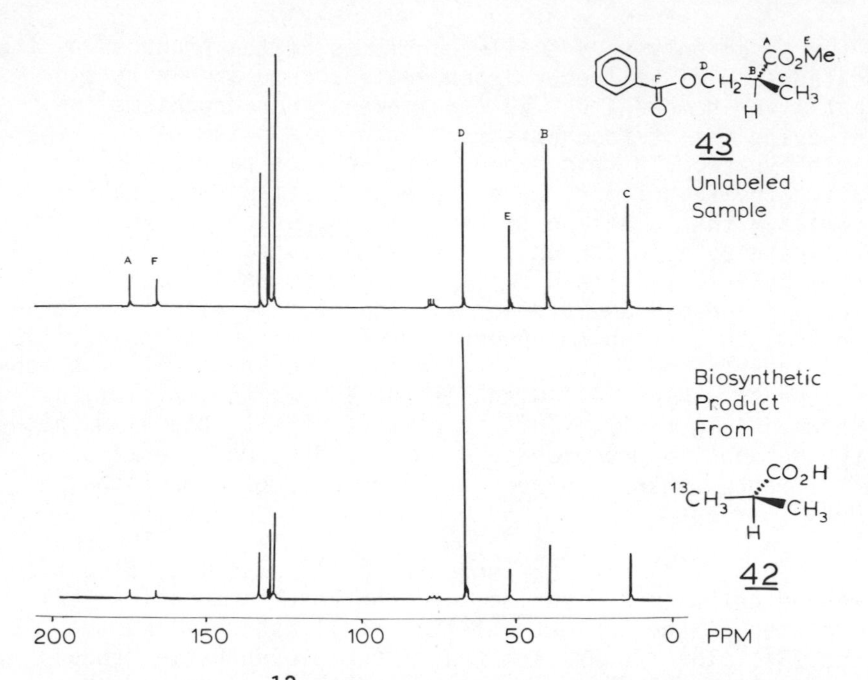

Figure 13. ^{13}C NMR spectra of β-HIBA methyl ester benzoate. Upper: unlabeled sample; lower: sample preparation from β-HIBA obtained by incubation of (2S)-[3-^{13}C] isobutyrate with Ps. putida.

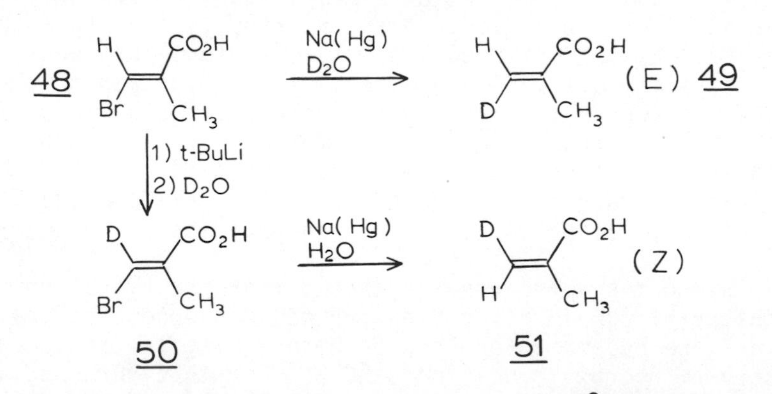

Figure 14. Synthesis of [E] and [Z]-[$^{3}H_6$]-methacrylic acids.

mine in some way the resultant chirality of the monodeuterated hydroxymethyl group. At the outset, it was established that the organism grew just as well on methacrylate as on isobutyrate.

The synthesis of stereospecifically deuterated methacrylic acid is outlined in Figure 14. The key step involved the Na(Hg) reduction of the vinylic bromide, 48, reported last year by Crout and Corkill[28]. At the time of their publication, we had been working along similar lines, and had found that bromide 48, on treatment with t-butyllithium, instead of undergoing lithium addition underwent vinyl hydrogen abstraction. Addition of D_2O then gave the deuterated bromo acid, 50, which by Crout's method[28] gave the [Z] deuterated methacrylic acid, 51.

With the availability of this labeled precursor, we turned our attention to development of a method useful for analyzing the chirality of a sample of β-HIBA methyl ester benzoate, 43, stereospecifically deuterated in the hydroxymethyl group. In the NMR spectrum of this compound (Figure 15) the signals for the diastereotopic C-3 hydrogens are too poorly resolved for analytical purposes. However, in the presence of $Eu(fod)_3$, these signals were shifted downfield to different extents, appearing in the 6-7 region as a well-resolved AB part of an ABX pattern. It thus became only necessary to assign the upfield and downfield halves of this AB pattern to the C-3 prochiral hydrogens of β-HIBA methyl ester benzoate.

The synthesis of a stereospecifically labeled reference sample of C-3 deuterated β-HIBA methyl ester benzoate is outlined in Figure 16. Treatment of [E]-[3-^{2}H]-methacrylic acid methyl ester, 52, with excess deuterated borane-THF gave the 2-methyl-1,3-butanediol, 53. We then protected one of the hydroxyl groups as the benzoate, a completely nonselective process. The resultant mixture of isotopic species 54 + 55 was then oxidized with permanganate to the acid, which was methylated with diazomethane. The product was a racemic mixture of two isotopic species of β-HIBA methyl ester benzoate, one (56) having no hydroxymethyl group hydrogens, the other (57) having a monodeuterated hydroxymethyl group of known configuration relative to the configuration at C-2. Since we were using a nonchiral shift reagent for C-3 hydrogen signal analysis, we expected both

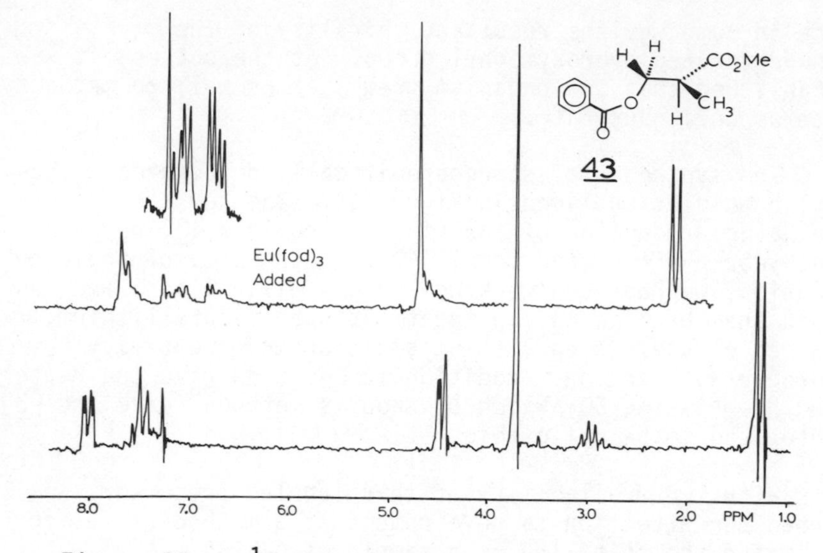

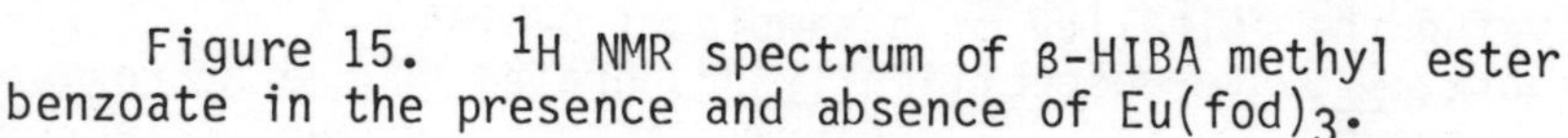

Figure 15. ^{1}H NMR spectrum of β-HIBA methyl ester benzoate in the presence and absence of $Eu(fod)_3$.

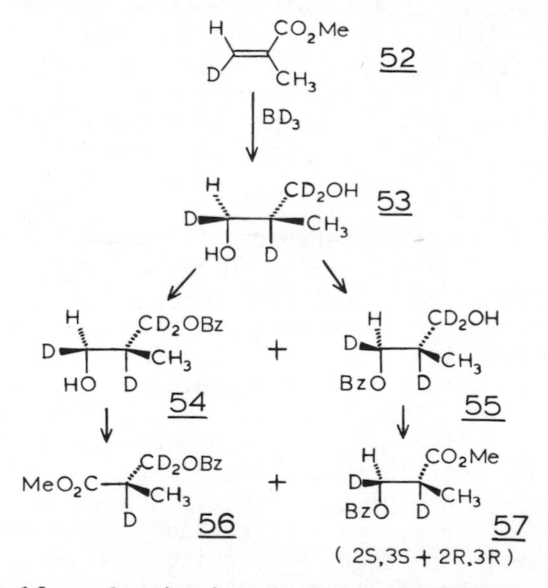

Figure 16. Synthesis of stereospecifically C-3 deuterated β-HIBA methyl ester benzoate for use in NMR signal assignment of C-3 hydrogens.

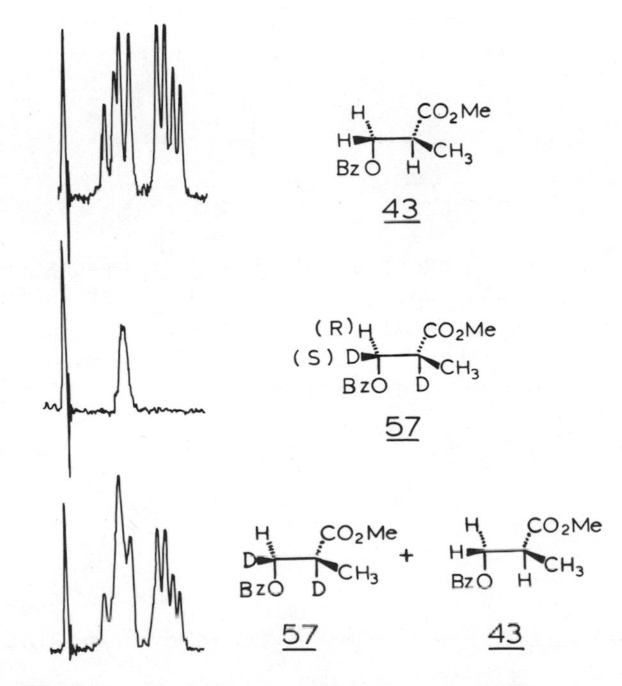

Figure 17. Low field region of $Eu(fod)_3$-shifted ^1H NMR spectra of (a) unlabeled β-HIBA methyl ester benzoate; (b) synthetic reference, (2S,3S + 2R,3R)-[3-$^2H_1$]-β-HIBA methyl ester benzoate; (c) mixture of (a) + (b).

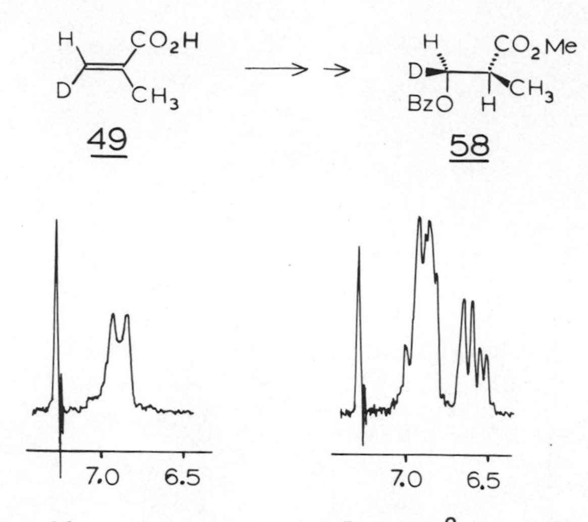

Figure 18. Conversion of [E]-[3-^{2}H]-methacrylic acid to β-HIBA by Ps. putida. Low field region of Eu(fod)$_3$-shifted ^{1}H NMR spectra of: left, biosynthetic β-HIBA methyl ester benzoate; right, biosynthetic product mixed with unlabeled β-HIBA methyl ester benzoate.

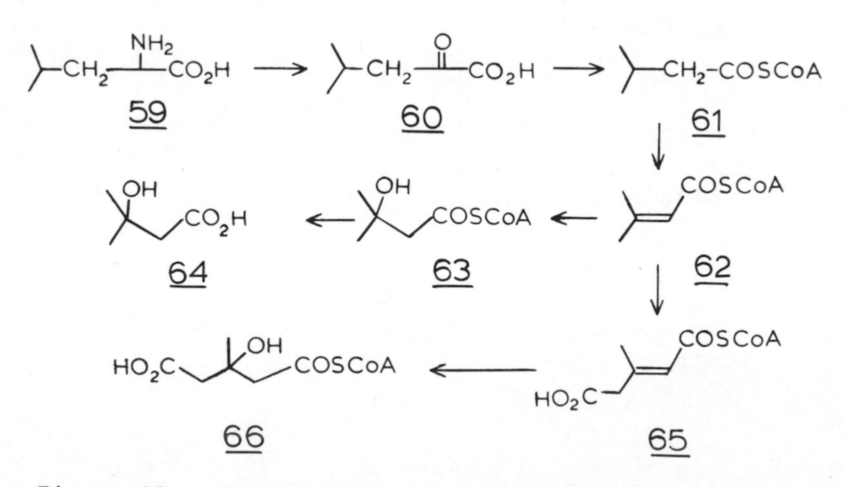

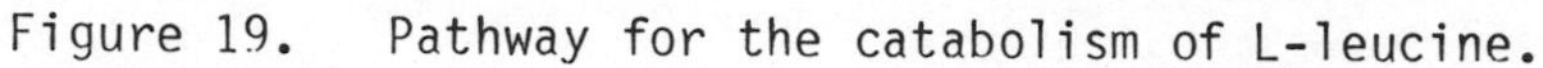
Figure 19. Pathway for the catabolism of L-leucine.

enantiomers to have identical NMR spectra (only the enantiomer of 57 having the C-2 configuration, the same as that of the metabolite of *Ps. putida* is shown in Figure 16).

The NMR spectrum of the synthetic reference 57 was then taken in the presence of $Eu(fod)_3$(Figure 17). As it was difficult to reproducibly shift the signal of interest to exactly the same position, after recording the spectrum of the labeled compound, an approximately equal amount of unlabeled β-HIBA methyl ester benzoate 43 plus additional $Eu(fod)_3$ were added, and the spectrum was again recorded. It can be seen that the downfield half of the AB pattern is due to the 3-*pro*-*R* hydrogen and the upfield half due to the 3-*pro*-*S* hydrogen, when C-2 has the *S*-configuration.

With the signal assignment secure, the biosynthetic experiment was performed. [E]-[3-^{2}H]-Methacrylic acid 49 was incubated with *Ps. putida*, and the β-HIBA isolated and converted to the methyl ester benzoate, 58. The $Eu(fod)_3$-shifted NMR spectrum (Figure 18) established that the deuterium is located in the 3-*pro*-*S* position. Therefore it may be concluded that the hydration proceeds by the *syn* addition of the elements of water to the *re*,*re* face of the carbon-carbon double bond of methacrylate. The stereochemistry of this process thus parallels that reported by Willadsen and Eggerer[29] for the dehydration of β-hydroxy-*n*-butyryl CoA by enoyl CoA hydratase, and by Sedgwick *et al*.[30] for the dehydration reaction catalyzed by yeast fatty acid synthetase.

Leucine catabolism. Having determined the stereochemistry of methacrylyl CoA hydration, we became interested in determining the stereochemistry of the hydration of βmethylcrotonyl CoA. This was of interest because it constitutes a step in the metabolism of *L*-leucine under certain circumstances. As will be seen, the results turned out to be rather surprising.

Figure 19 shows the catabolic pathway of *L*-leucine[31]. It should be noted that the intermediate β-methylcrotonyl CoA (62) is normally first carboxylated, by β-methylcrotonyl CoA carboxylase[32], a biotin-dependent enzyme, to β-methylglutaconyl CoA (65). This is then hydrated by the enzyme methylglutaconase to β-hydroxy-β-methylglutaryl CoA (66) which can undergo fragmentation to acetoacetate, or be reduced to mevalonic acid. In rats having induced biotin-

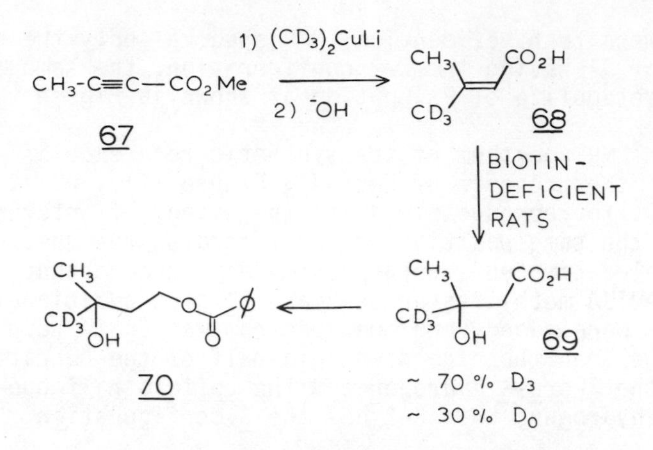

Figure 20. Synthesis of (E)-[3-$^2H_3$]-β-methylcrotonic acid and its conversion by biotin-deficient rats to β-hydroxyisovaleric acid. Conversion of the biosynthetic product via methylation (CH_2N_2), reduction ($LiAlH_4$) and benzoylation to (3-$^2H_3$)-3-methyl-1,3-butanediol 1-benzoate.

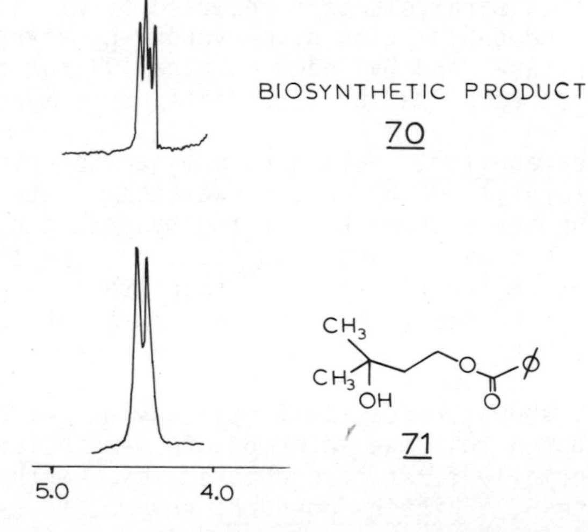

Figure 21. Methyl region of Eu(hfbc)$_3$-shifted NMR spectra of 3-methyl-1,3-butanediol 1-benzoate. Lower: unlabeled sample; upper: biosynthetic product consisting of ca. 65-70% 3-$^2H_3$ and 30-35% unlabeled material.

deficiency[33], or in humans having the inborn error of leucine metabolism, β-hydroxyisovaleric aciduria[34], the carboxylation of β-methylcrotonyl CoA is defective and hydration of β-methylcrotonyl CoA occurs instead, leading to the accumulation of β-hydroxyisovaleric acid 64 in the urine.

For this study, (E)-[methyl-$^2H_3$]-β-methylcrotonic acid, 8, was prepared as shown in Figure 20. (The method is similar to that used by Hill and coworkers[10] for the synthesis of the (Z) isomer of this compound). Biotin-deficient rats were obtained by feeding weanling rats a diet consisting of 30% uncooked egg white as the sole source of protein for 7-8 weeks. After this period, the rats showed signs of severe biotin deficiency (extreme dermatitis, fur loss, weakness, etc). The (E)-[methyl-$^2H_3$]-β-methylcrotonate 68 was then administered as the sodium salt (0.2 M in saline) by intraperitoneal injection, in doses of 50 mg/day/rat to two rats. Each rat excreted in the urine ca. 15-20 mg/day of β-hydroxyisovaleric acid, 69, consisting of approximately 65-70% C^2H_3 labeled and 30-35% unlabeled material (resulting from leucine in the diet, or other sources).

For determination of the resulting labeling stereochemistry of the biosynthetic product, the extracted β-hydroxyisovaleric acid was converted to the methyl ester with diazomethane, and reduced to the diol with $LiAlH_4$. The product was then converted to the monobenzoate, 70 (Figure 20). The NMR spectrum of this benzoate was then taken in the presence of tris-[3-(heptafluorobutyryl)-d camphorato] europium(III) [Eu(hfbc)$_3$]. We had established in advance that this chiral shift reagent caused a separation of signals of the enantiotopic methyls of this benzoate.

Figure 21 shows the shifted NMR signals of the methyls of the unlabeled benzoate, 71, and of the biosynthetic product, 70. While the unlabeled sample showed two methyl signals, the biosynthetic product showed four methyl signals. Apparently the shift reagent complexes to a slightly different extent with the deuterated material than with the unlabeled compound. Clearly, however, since four methyl signals (of roughly equal intensities) were visible in the NMR of the biosynthetic product mixed with unlabeled material, the biosynthetic product was not stereospecifically labeled but consisted of an equal mixture of two enantiomers.

The stereochemistry of hydrogen addition at C-2 in this reaction has not yet been determined. However, the stereochemical result for C-3 contrasts sharply with that observed in the previously discussed cases. The reaction must be enzyme-catalyzed, since β-methylcrotonyl CoA does not hydrate spontaneously. The results suggest that the active site of the enzyme is restricted with respect to accessibility of the C-3 position of the substrate. While it can accommodate one methyl at C-3, it apparently cannot readily accommodate two such groups. Hence, hydroxyl addition is not controlled, and occurs to an equal extent on both faces of the double bond.

ACKNOWLEDGMENTS

The author wishes to express his appreciation to his colleagues, Dr. Lawrence J. Lin, Dr. Yeou-Ruoh Chu, and Mr. Chou-Hong Tann for their contributions to the above studies, to Dr. Norbert Neuss and his colleagues at Eli Lilly Research Laboratories, Indianapolis for carrying out the conversions of labeled valines into penicillins; to the National Institutes of Health for support of the studies; and to Dr. K. Tanaka, Yale University, for helpful discussions.

1. Arnstein, H.R.V. and D. Morris. 1960. The Structure of a Peptide Containing α-Aminoadipic Acid, Cystine, and Valine, Present in the Mycelium of Penicillin chrysogenum. Biochem. J. 76:357-361.
2. Loder, P.B. and E.P. Abraham. 1971. Isolation and Nature of Intracellular Peptides from a Cephalosporin C-Producing Cephalosporium sp. Biochem. J. 123:471-476.
3. Fawcett, P. A., P. B. Loder, M. J. Duncan, T. J. Beesley, and E. P. Abraham. 1973. Formation and Properties of Protoplasts from Antibiotic-producing Strains of Penicillium chrysogenum and Cephalosporium acremonium. J. Gen. Microbiol. 79:293-309.
4. Aberhart, D. J. 1977. Biosynthesis of β-Lactam Antibiotics. Tetrahedron, 33:1545-1559.
5. Cooper, R. D. G., and D. O. Spry. 1972. Rearrangements of Cephalosporins and Penicillins. In Cephalosporins and Penicillins, Chemistry and Biology, E. H. Flynn, ed. Academic Press, p. 183-254.
6. Aberhart, D. J., J. Y. R. Chu, N. Nuess, C. H. Nash, J. Occolowitz, L. L. Huckstep, and N. DeLa Higuera. 1974. Retention of Valine Methyl Hydrogens in Penicillin Biosynthesis. J. Chem. Soc., Chem. Comm.:564-565.
7. Archer, R. A., R. D. G. Cooper, and P. V. Demarco. 1970. Structural Studies on Penicillin Derivatives: ^{13}C Nuclear Magnetic Resonance Studies of Some Penicillins and Related Sulfoxides. J. Chem. Soc., Chem. Comm.:1291-1293.
8. Aberhart, D. J. and L. J. Lin. 1973. Studies on the Biosynthesis of β-Lactam Antibiotics, I. Synthesis of (2RS,3S)-[4,4,4-$^{2}H_3$] Valine. J. Am. Chem. Soc., 95:7859-7860.
9. Aberhart, D. J. and L. J. Lin. 1974. Studies on the Biosynthesis of β-Lactam Antibiotics, Part 1. Stereospecific Synthesis of (2RS,3S)-[4,4,4-$^{2}H_3$]-,(2RS,3S)-[4-^{3}H]-, (3RS, 3R)-[4-^{3}H]-, and (2RS,3S)-[4-^{13}C]-Valine. Incorporation of (2RS,3S)-4-^{13}C]-Valine into Penicillin V. J. Chem. Soc., Perkin Trans. I:2320-2326.
10. Hill, R. K., S. Yan, and S. M. Arfin. 1973. Enzymatic Discrimination between Diastereotopic Enol Faces in the Dehydrase Step of Valine Biosynthesis. J. Am. Chem. Soc. 95:7857-7859.

11. Baldwin, J. E., J. Loliger, W. Rastetter, N. Nuess, L. L. Huckstep, and N. DeLa Higuera. 1973. Use of Chiral Isopropl Groups in Biosynthesis. Synthesis of (2RS,3S)-[4-^{13}C] Valine. J. Am. Chem. Soc. 95:3796-3797.
12. Nuess, N., C. H. Nash, J. E. Baldwin, P. A. Lemke, and J. B. Grutzner. 1973. Incorporation of (2RS,3S)-[4-^{13}C] Valine into Cephalosporin C. J. Am. Chem. Soc. 95:3797-3798.
13. Kluender, H., C. H. Bradley, C. J. Sih, P. Fawcett, and E. P. Abraham. 1973. Synthesis and Incorporation of (2RS,3S)-[4-^{13}C] Valine into β-Lactam Antibiotics. J. Am. Chem. Soc. 95:6149-6150.
14. Aberhart, D. J., L. J. Lin, and J. Y. R. Chu. 1975. Studies on the Biosynthesis of β-Lactam Antibiotics. Part II. Synthesis, and Incorporation into Penicillin G, of (2RS,2'RS,3R,3'R)-[3,3'-$^{3}H_2$]-Cystine and (2RS,2'RS,3S,3'S)-[3,3'-$^{3}H_2$]-Cystine. J. Chem. Soc., Perkin Trans. I:2517-2523.
15. Huddleston, J. A., E. P. Abraham, D. W. Young, D. J. Morecambe, and P. K. Sen. 1978. The Stereochemistry of β-Lactam Formation in Cephalosporin Biosynthesis. Biochem. J. 169:705-707.
16. Morecambe, D. J. and D. W. Young. 1975. Synthesis of Chirally Labelled Cysteines and the Steric Origin of C(5) in Penicillin Biosynthesis. J. Chem. Soc. Chem. Comm.:198.
17. Young, D. W., D. J. Morecambe, and P. K. Sen. 1977. The Stereochemistry of β-Lactam Formation in Penicillin Biosynthesis. Eur. J. Biochem. 75:133.
18a. Robinson, W. G., R. Nagle, B. K. Bachhawat, F. P. Kupiecki, and M. J. Coon. 1957. Coenzyme A Thiol Esters of Isobutyric, Methacrylic, and β-Hydroxyisobutyric Acids as Intermediates in the Enzymatic Degradation of Valine. J. Biol. Chem. 224:1-9.
18b. Robinson, W. G., and M. J. Coon. 1957. The Purification and Properties of β-Hydroxyisobutyric Dehydrogenase. J. Biol. Chem. 225:511-519.
19. Baretz, B. and K. Tanaka. 1978. Metabolism in Rats in Vivo of Isobutyrates Labeled with Stable Isotopes at Various Positions. Identification of Propionate as an Obligate Intermediate. J. Biol. Chem. 253:4203-4213.
20. Tanaka, K. and I. M. Armitage. 1975. Investigaion of ^{13}C-Valine Metabolism in Methylmalonic

Acidemia Using Nuclear Magnetic Resonance: Identification of Propionate as an Obligate Intermediate. From Proc. Second Interna Conf. Stable Isotopes. E. R. Klein and P. D. Klein, eds. National Technical Information Service, U.S. Department of Commerce, Springfield, Virginia.

21. Tanaka, K., I. M. Armitage, H. S. Ramsdell, Y. E. Hsia, S. R. Lipsky, and L. E. Rosenburg. 1975. [^{13}C]-Valine Metabolism in Methylmalonic-acidemia Using Nuclear Magnetic Resonance: Propionate as an Obligate Intermediate. Proc. Nat. Acad. Sci. USA 72:3692-3696.
22. Marshall, V. P., and J. R. Sokatch. 1972. Regulation of Valine Catabolism in Pseudomonas putida. J. Bacteriol. 110:1073-1081.
23. Tanaka, K. Personal communication to D.J. Aberhart.
24. Aberhart, D. J. 1975. Synthesis of (2R)-[3,3,3-d_3] Isobutyric Acid, Ammonium Salt. Tetrahedron Lett:4373-4374.
25. Clarke, T. G., N. A. Hampton, J. B. Lee, J. R. Morley, and B. Scanlon. 1970. Oxidations Involving Silver. Part VI. Oxidation of α-Amino-acids and α-Amino-esters with Silver(II) Picolinate and Silver(II) Oxide. J. Chem. Soc. (C): 815-817.
26. Goodhue, C. T. and J. R. Schaeffer. 1971. Preparation of L(+)-β-Hydroxyisobutyric Acid by Bacterial Oxidation of Isobutyric Acid. Biotechnol. Bioeng. 13:203-214.
27. Aberhart, D. J. 1977. A Stereochemical Study on the Metabolism of Isobutyrate in Pseudomonas putida. Bioorg. Chem. 6: 191-201.
28. Crout, D. H. G. and J. A. Corkill. 1977. Sodium Amalgam Reduction of 3-Bromopropenic Acids: A Convenient Stereoscopic Synthesis of [3-$^3H_1$] and [3-$^2H_1$] Acrylic Acids. Tetrahedron Lett: 4355-4357.
29. Willadsen, P. and H. Eggerer. 1975. Substrate Stereochemistry of the Enoyl-CoA Hydratase Reaction. Eur. J. Biochem. 54:247-252.
30. Sedgwick, B., C. Morris, and S. J. French. 1978. Stereochemistry Course of Dehydration Catalysed by the Yeast Fatty Acid Synthetase. J. Chem. Soc., Chem. Comm:193-194.
31. Meister, A. 1965. Biochemistry of the Amino Acids II:729-753. Academic Press, New York.

32. Del Campillo-Campbell, A., E. E. Dekker, and M. J. Coon. 1959. Carboxylation of β-Methylcrotonyl Coenzyme A by a Purified Enzyme from Chicken Liver. Biochem. Biophys. Acta 31:290-292.
33. Tanaka, K. and K. J. Isselbacher. 1970. Experimental β-Hydroxyisovaleric Aciduria Induced by Biotin Deficiency. Lancet October 31:930-931.
34. Eldjarn, L., E. Jellum, O. Stokke, H. Pande, and P. E. Waaler. 1970. β-Hydroxyisovaleric Acidurea and β-Methylacrotonylglycinuria: A New Inborn Error of Metabolism. Lancet September 5: 521-522.

Chapter Three

BIOSYNTHESIS OF THE CEPHALOTAXUS ALKALOIDS

RONALD J. PARRY

Department of Chemistry
Rice University
Houston, Texas

INTRODUCTION

Conifers of the genus *Cephalotaxus* are members of a monogeneric family of plants (Cephalotaxaceae) that occurs in the Far East. These plants have been found to contain a group of alkaloids of unique structure. The most abundant member of this group is cephalotaxine (1) (Figure 1), which was first isolated[1] in 1963. The structure of cephalotaxine was elucidated[2] in 1969 by X-ray analysis of the methiodine. In *C. harringtonia*, cephalotaxine is accompanied by small quantities of related alkaloids[3-5], the most important being harringtonine (2), deoxyharringtonine (3), isoharringtonine (4), and homoharringtonine (5) (Figure 1). Each of these four alkaloids possesses significant inhibitory activity against the experimental lymphoid leukemia systems P388 and L-1210 in mice[6]. In addition to the alkaloids based upon the "cephalotaxane" skeleton, a number of alkaloids containing the "homoerythrina" skeleton occur in *C. harringtonia*[7]. One such compound is 3-epischelhammericine (6)

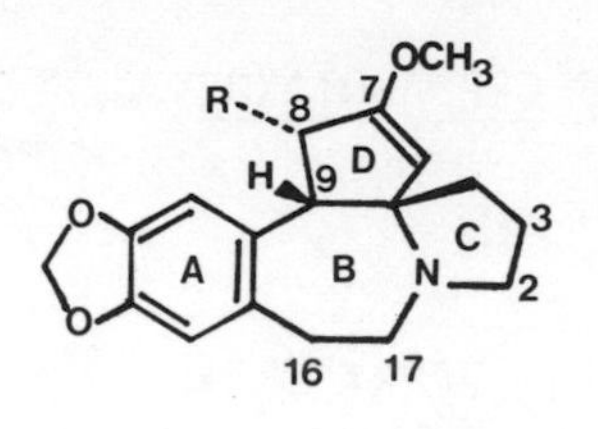

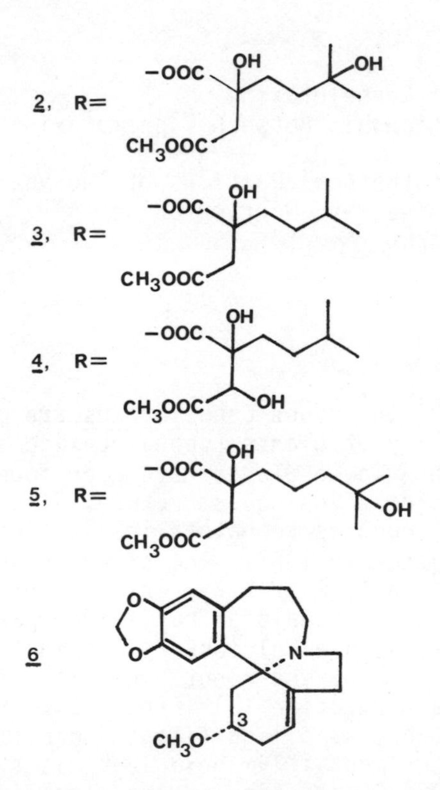

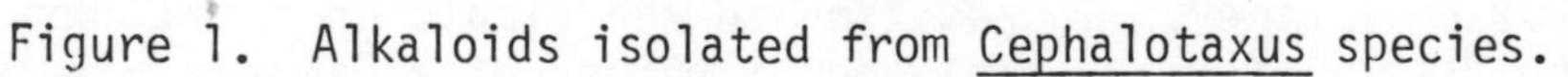

Figure 1. Alkaloids isolated from Cephalotaxus species.

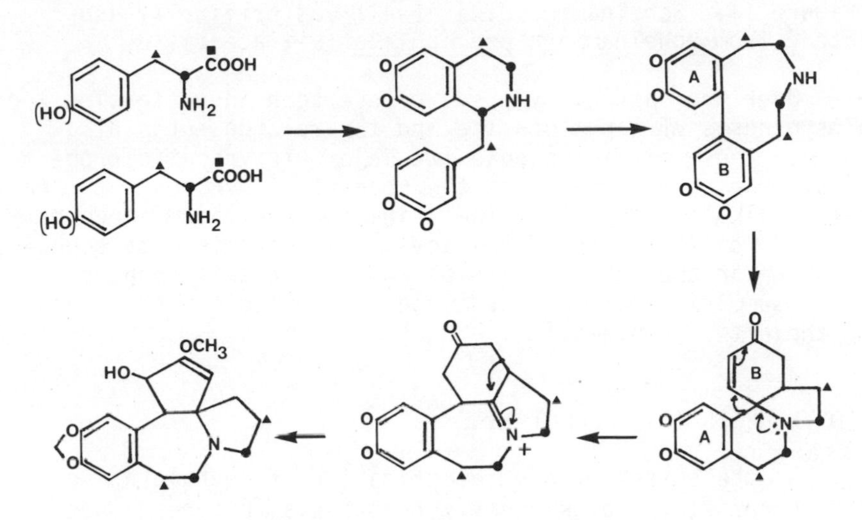

Figure 2. Hypothetical biosynthesis of cephalotaxine via a 1-benzyltetrahydroisoquinoline.

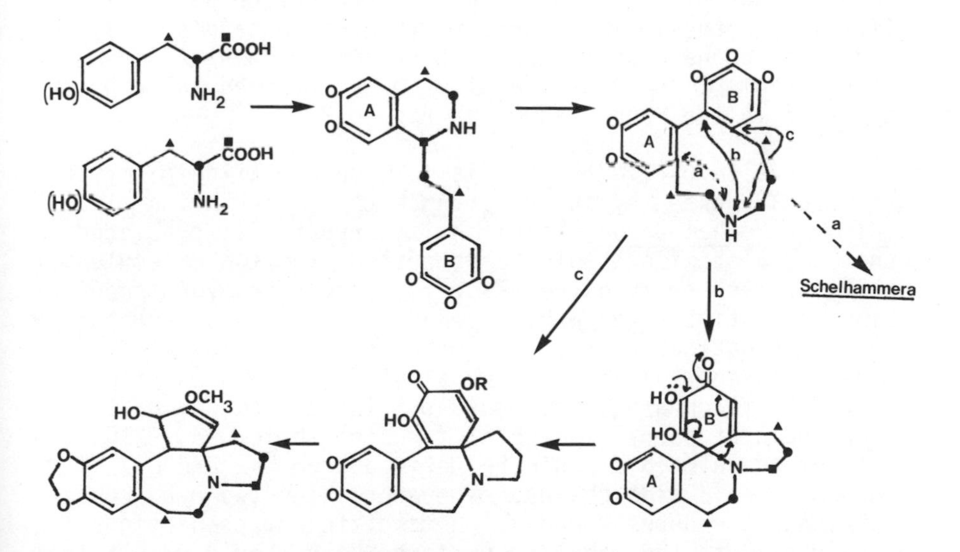

Figure 3. Hypothetical biosynthesis of cephalotaxine via a 1-phenethyltetrahydroisoquinoline.

(Figure 1). Schelhammericine itself was originally isolated[8] from Schelhammera pedunculata (Liliaceae).

Over the last six years, we have been investigating the biosynthesis of cephalotaxine and the related ester alkaloids. These alkaloids pose two major biosynthetic problems. One problem concerns the mode of biosynthesis of the parent alkaloid cephalotaxine. The other problem concerns the mode of formation of the acyl residues linked to cephalotaxine in the ester alkaloids 2 - 5. In this chapter I shall summarize our current state of knowledge with regard to these two problems.

BIOSYNTHESIS OF CEPHALOTAXINE

At the outset of our investigation of cephalotaxine biosynthesis, two biosynthetic hypothesis (Figure 2) was derived from Barton's investigations[9,10] of the biosynthesis of the alkaloids of Erythrina plants (Fabaceae). This hypothesis predicted that cephalotaxine should be formed from two molecules of phenylalanine or tyrosine via the intermediacy of a benzyltetrahydroisoquinoline derivative. If this hypothesis were correct, then carbon atoms 2, 3, 16, and 17 of cephalotaxine were expected to be derived from carbon atoms 2 and 3 of the side chains of prenylalanine or tyrosine in the manner shown in Figure 2.

The alternative hypothesis for cephalotaxine biosynthesis was suggested by the occurrence of homoerythrina alkaloids in Cephalotaxus plants[7]. This hypothesis predicted that cephalotaxine should be generated from two molecules of phenylalanine or tyrosine via a phenethyltetrahydroisoquinoline derivative (Figure 3). Because of the results obtained in investigations of the biosynthesis of colchicine[11], a modified phenethylisoquinoline alkaloid, it was anticipated that C-2 of cephalotaxine would originate from C-1 of the side chain of phenylalanine or tyrosine while C-3, C-16, and C-17 of the alkaloid would be derived from C-2 and C-3 of the amino acid side chains. Therefore, the two hypotheses outlined in Figures 2 and 3 led to different predictions with regard to the labeling pattern which should result from the incorporation of side chain-labeled phenylalanine or tyrosine into cephalotaxine.

Table 1. Incorporation of Precursors into Cephalotaxine by C. harringtonia var. fastigiata

Expt. No.	Precursor	Feeding Period (weeks)	% Incorporation	Distribution of ^{14}C-Activity in Cephalotaxine
1	[2-^{14}C]-DL-Tyrosine	8	0.035	37% at C-17, 0% at C-2, C-3
2	[2-^{14}C]-DL-Tyrosine	8	0.16	37% at C-17, 0% at C-2, C-3
3	[ring-^{14}C]-L-Tyrosine	8	0.11	90% in C-10 to C-15, C-18
4	[p-^{14}C]-L-Tyrosine	8	0.06	93% in C-9 to C-16, C-18
5	[2-^{14}C]-DL-Tyrosine	2	0.04	49% at C-17
6	[2-^{14}C]-DL-Tyrosine	1	0.003	63% at C-17
7	12-^{14}C]-DL-Phenylalanine	2	0.02	84% at C-2
8	[1-^{14}C]-Cinnamic acid	2	0.004	9% at C-2
9	[1-^{14}C]-DL-Tyrosine	2	0.006	3% at C-2
10	3(RS)-[3-^{3}H-2-^{14}C]-DL-Phenylalanine, ^{3}H:^{14}C = 4.55	2	0.1	^{3}H:^{14}C = 2.12
11.	3(RS)-[3-^{3}H-p-^{14}C]-DL-Phenylalanine, ^{3}H:^{14}C = 3.48	2	0.005	^{3}H:^{14}C + 1.60, 100% at C-8

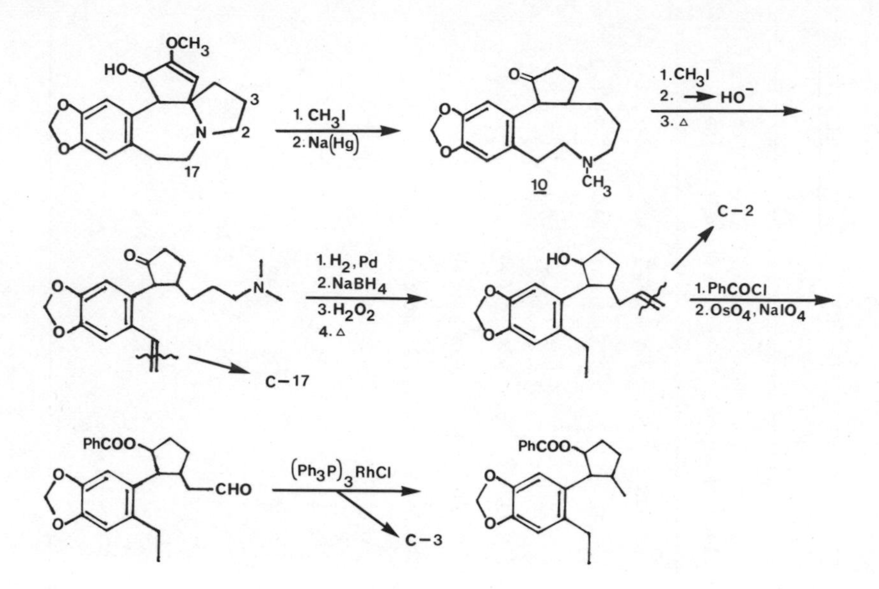

Figure 4. Degradation of cephalotaxine.

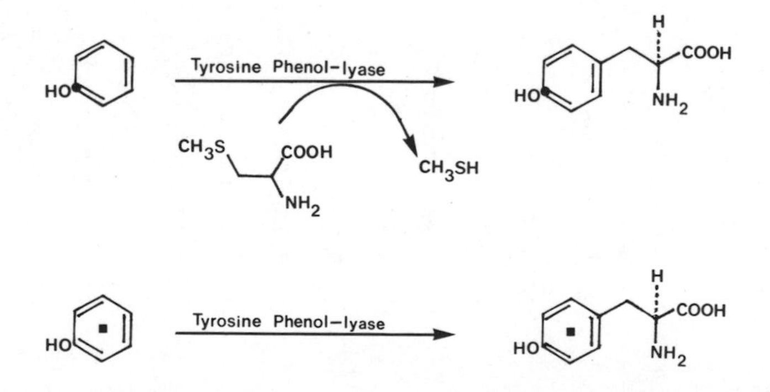

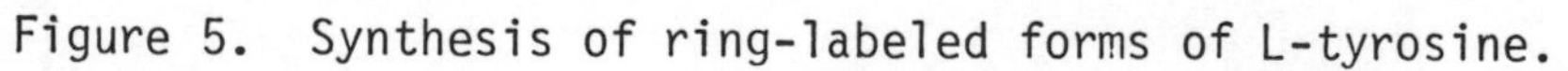

Figure 5. Synthesis of ring-labeled forms of L-tyrosine.

In 1972 preliminary incorporation experiments with labeled tyrosine and phenylalanine were carried out using *C. harringtonia* var. *fastigiata* growing in a greenhouse. Precursors were administered to cuttings and to whole plants via the cotton wick method. The use of cuttings proved unsatisfactory. Short incorporation periods (< 7 days) using intact plants gave insignificant incorporations of phenylalanine and low incorporations of tyrosine. Increasing the length of the incorporation period led to increasing levels of tyrosine incorporation. Table 1, expt. 1, shows a typical result: administration of [2-^{14}C]-DL-tyrosine for an 8-week period yielded radioactive cephalotaxine with an incorporation of 0.035%. Additional experiments with [2-^{14}C]-DL-tyrosine were then carried out using rapidly growing *C. harringtonia* plants living in an environmental chamber. Under these conditions significantly higher incorporations of tyrosine were observed after an 8-week period (Table 1, expt. 2). Degradation of the radioactive cephalotaxine in the manner outlined in Figure 4 provided access to C-17, C-2, and C-3 of the alkaloid. The two samples of radioactive cephalotaxine obtained by administration of [2-^{14}C]-DL-tyrosine to *Cephalotaxus* growing either in a greenhouse or in an environmental chamber gave identical results[12]: no significant amount of radioactivity was present at C-2 or C-3 of the alkaloid, and, after correction for the generation of radioinactive formaldehyde from *N*-methyl groups, 37% of the total radioactivity was present at C-17. These observations were surprising since the labeling pattern did not correspond to that expected on the basis of either of the biosynthetic hypotheses shown in Figures 1 and 2. Further degradative work failed to locate the remaining 63% of the radioactivity present in the labeled cephalotaxine produced in experiments 1 and 2.

An explanation for the unusual labeling pattern observed in experiments 1 and 2 was provided by additional investigations. Samples of [*p*-^{14}C]- and [ring-^{14}C]-L-tyrosine were synthesized according to the method outlined in Figure 5. Administration of these forms of labeled tyrosine to *C. harringtonia* under the usual conditions yielded radioactive cephalotaxine which was then degradations (Table 1, expts. 3,4) (Figure 7) established that the ring-labeled tyrosines lead to exclusive labeling of ring A of cephalotaxine under the same conditions which lead to 37% incorporation of the label of [2-^{14}C]-tyrosine into C-17 of the

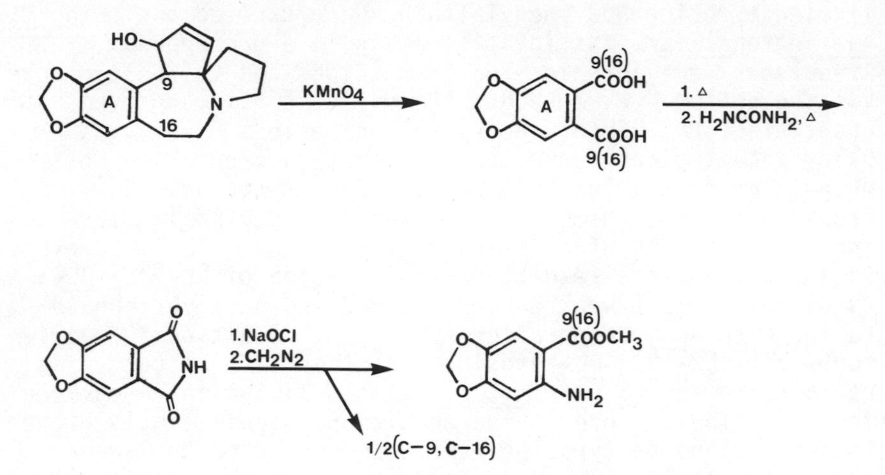

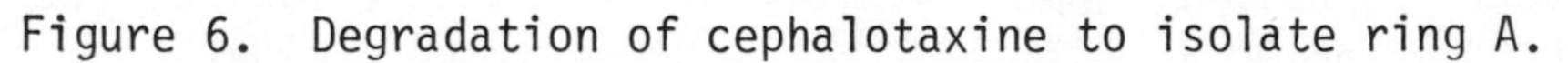
Figure 6. Degradation of cephalotaxine to isolate ring A.

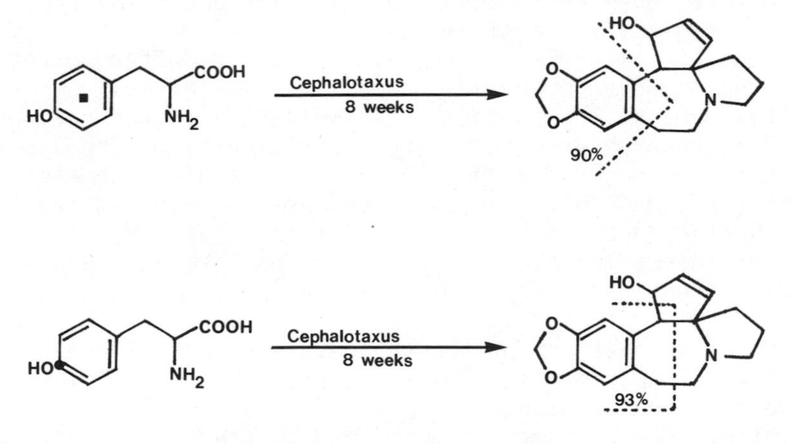

Figure 7. Degradative results from incorporation of ring-labeled tyrosines into cephalotaxine.

alkaloid[13]. The non-exclusive incorporation of label into C-17 could therefore be explained to be the result of catabolism of the tyrosine, possibly by enzyme-catalyzed cleavage to phenol and serine followed by incorporation of the latter substance into cephalotaxine. Evidence favoring such an explanation was obtained by administration of [2-^{14}C]-tyrosine to *Cephalotaxus* plants for periods of decreasing duration (Table 1, expts. 5,6). The results of these experiments were gratifying: degradation of the radioactive cephalotaxine revealed an *increasing* specificity of labeling at C-17 as the incorporation period was decreased[13].

The experiments just outlined provided compelling evidence for the derivation of ring A and carbon atoms 16 and 17 of cephalotaxine from the amino acid tyrosine. This conclusion was also supported by experiments involving administration of [3-^{14}C]-*DL*-tyrosine to *Cephalotaxus* for an 8-week period, but in these experiments also, some catabolism of the labeled tyrosine appeared to be taking place[12].

The clarification of the role of tyrosine in the biosynthesis of cephalotaxine raised the question of the origin of the remaining carbon atoms of the cephalotaxine skeleton. Although our initial incorporation experiments with phenylalanine had been unpromising, the analogy provided by colchicine required that we reexamine the potential of phenylalanine as a cephalotaxine precursor. An additional impetus was provided by a report from Battersby's laboratory[14] that the biosynthesis of schelhammeridine in *Schelhammera* plants proceeded in accordance with the colchicine analogy. In the event, administration of 1-^{14}C-*DL*-phenylalanine to *C. harringtonia* plants growing in an environmental chamber yielded radioactive cephalotaxine after a 2-week period (Table 1, expt. 7). Degradation of the alkaloid according to the route outlined in Figure 4 revealed that 84% of the total radioactivity was present at C-2[13]. This observation provided substantial evidence that cephalotaxine is a member of the family of phenethylisoquinoline alkaloids that includes the *Schelhammera* and *Colchicum* alkaloids and it suggested that ring D of cephalotaxine is derived from the aromatic ring of phenylalanine by the loss of one carbon atom.

Since it had been established that phenylalanine is incorporated into colchicine via cinnamic acid[11], the incorporation of $[1-^{14}C]$-cinnamic acid into cephalotaxine was investigated. The results of two feeding experiments are shown in Table 1, expts. 8 and 9. Both experiments yielded radioactive alkaloid, but, to our surprise, very little of the radioactivity was present at C-2. The apparent inability of cinnamate to serve as a specific precursor of cephalotaxine raised the question of the nature of the steps associated with the loss of the nature of the steps associated with the loss of the amino group from the side chain of phenylalanine during its conversion to cephalotaxine. If the amino group were removed by elimination of ammonia, as occurs in the conversion of phenylalanine to cinnamic acid[15], then a stereospecific loss of one hydrogen atom from C-3 of the amino acid should take place[15]. Administration of 3(RS)-$[^{3}H-2-^{14}C]$-DL-phenylalanine to *Cephalotaxus* demonstrated that such a process indeed occurs (Table 1, expt. 10)[13]: the tritium to carbon-14 ratio in the radioactive alkaloid corresponded to a 54% loss of tritium (expected loss is 50%). This result suggests that an intermediate with α,β-unsaturation lies on the pathweay between phenylalanine and cephalotaxine. The intermediate may in fact be cinnamic acid, in which case our failure to observe specific incorporation of cinnamate into cephalotaxine could be attributed to an inability of the acid to reach the site of alkaloid biosynthesis.

Alternatively, the loss of tritium observed in experiment 10 could be due to the operation of an exchange process catalyzed by the enzyme tautomerase[16], rather than to a loss of ammonia via an ammonia lyase reaction. These two possibilities should lead to different stereochemical results with regard to hydrogen removal from C-3: the ammonia lyase reaction should proceed with removal of the 3-*pro*-*S* hydrogen atom[15], while the tautomerase reaction should cause loss of the 3-*pro*-*R* hydrogen atom[16]. In order to examine this point, samples of 3(R)-^{3}H]- and 3(*S*)-$[^{3}H]$-*DL*-phenylalanine were administered to *Cephalotaxus* in conjunction with $[2-^{14}C]$-*DL*-phenylalanine. These incorporation experiments were carried out under conditions which appeared to be identical to those employed in experiments 7 and 10 (Table 1); nevertheless, negligible incorporation of the doubly labeled phenylalanine into cephalotaxine was observed in each case! The failure of phenylalanine to be incorporated into cephalotaxine in these experiments suggests that there are some unknown variables which strongly influence the efficiency of incorporation of this amino acid into cephalotaxine. A possible clue as to the nature of one such variable may have been found and it will be referred to in the discussion to follow.

One of the more intriguing features of cephalotaxine biosynthesis is the loss of one carbon atom from the aromatic ring of phenylalanine as a result of formation of ring D of the alkaloid. Some insight into the nature of the carbon excision process has been obtained with the aid of [p-^{14}C]-phenylalanine. The biosynthetic pathway outlined in Figure 3 suggests that administration of [p-^{14}C]-phenylalanine to Cephalotaxus should lead to a labeled 1-phenethyltetrahydroisoquinoline such as 7 (Figure 8) which would be transformed into a labeled diketone (8) or triketone (9) via oxidative phenolic coupling. The loss of one carbon atom from 8 or 9 could then be visualized as the result of a benzilic acid rearrangement[7]. Such a rearrangement could proceed in three ways (Figure 9). One pathway (path a) would lead to loss of the para carbon atom of phenylalanine. The second pathway (path b) would lead to the loss of one of the meta carbon atoms with resultant labeling of the cephalotaxine at C-8. The third pathway (path c) would proceed with loss of the alternative meta carbon atom and produce cephalotaxine labeled at C-7. Thus, it should be possible in principle to distinguish between these three pathways.

[p-^{14}C]-Aniline was prepared by modification of literature syntheses of ring-labeled benzene derivatives[17,18] (Figure 10). [p-^{14}C]-DL-Phenylalanine was then synthesized from the labeled analine via the Meerwein arylation reaction (Figure 10). The ring-labeled phenylalanine was mixed with 3(RS)-[^{3}H]-DL-phenylalanine so that tritium would serve as a reference label in the event that the para carbon atom was lost. The doubly labeled precursor having been prepared, we experienced some apprehension with regard to the successful outcome of the incorporation experiment. This apprehension was due to the fact that our two prior attempts to incorporate phenylalanine into cephalotaxine had failed. A successful incorporation experiment was achieved in the present case by utilizing plants that had been removed from an environmental chamber programmed for a short-day cycle and placed in a chamber with a long-day cycle. After about two weeks these plants began to grow vigorously and they were then administered the precursor. It is interesting to note

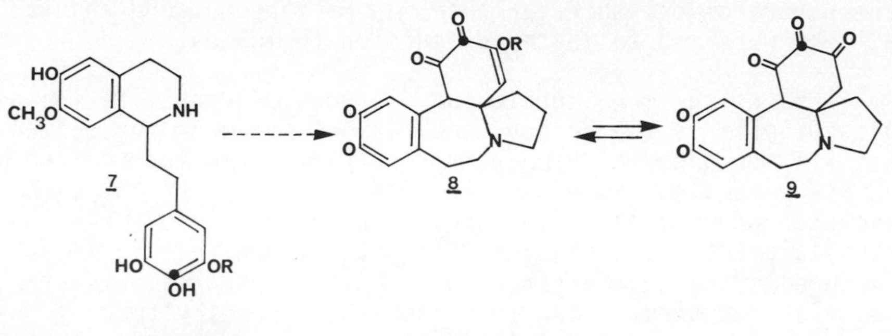

Figure 8. Hypothetical intermediates in cephalotaxine biosynthesis from [p-^{14}C]-phenylalanine.

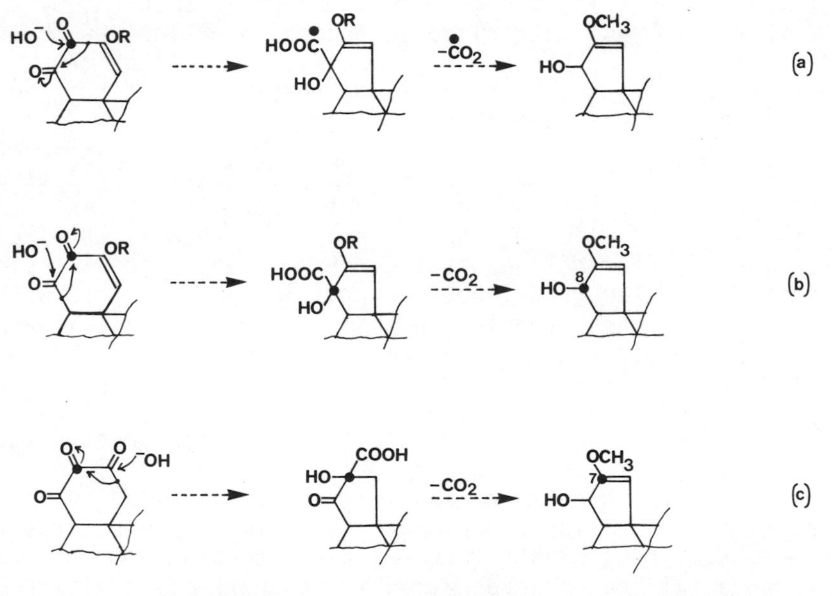

Figure 9. Hypothetical mechanisms for formation of ring D of cephalotaxine.

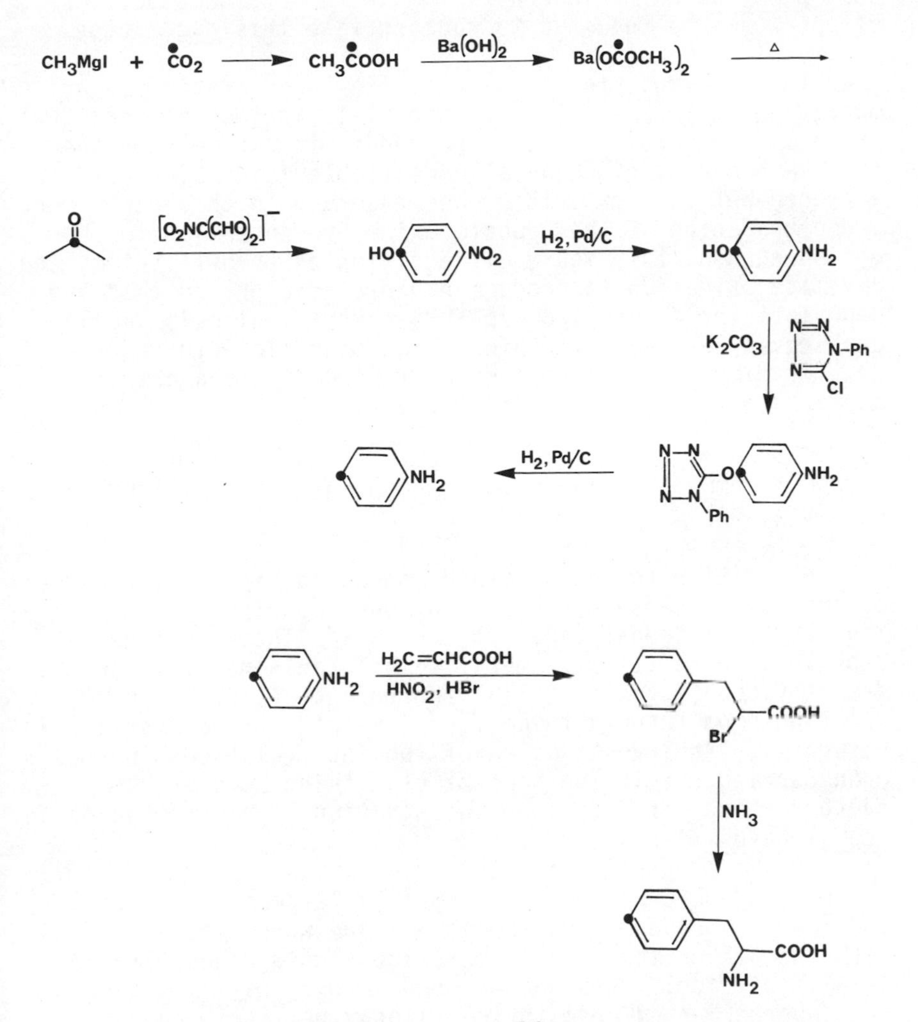

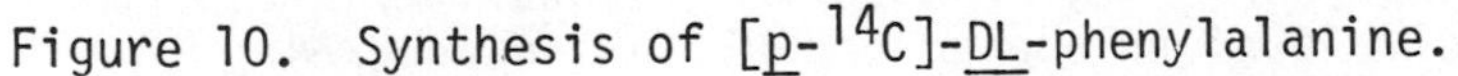
Figure 10. Synthesis of [p-^{14}C]-DL-phenylalanine.

that the plants used in the two experiments where phenylalanine incorporation had failed had been growing on a long-day cycle for many months. This may have been the reason why poor phenylkalanine incorporations were obtained, but more effort would be required to substantiate this suggestion.

In any event, the results of the incorporation experiment with 3(RS)-[p-^{14}C-^{3}H]-DL-phenylalanine established that the ring contraction process proceeds without loss of the para carbon atom of phenylalanine (Table 1, expt. 11). The expected 50% loss of tritium was observed in this experiment and degradation of the cephalotaxine by conversion to the cyclopentanone 10 (Figure 4), addition of phenyllithium, and chromate oxidation to produce benzoic acid proved that the cephalotaxine formed from [p-^{14}C]-phenylalanine is labeled exclusively at C-8. Therefore, the benzilic acid rearrangement leading to the formation of ring D of cephalotaxine appears to follow path b in Figure 9.

METABOLIC RELATIONSHIPS BETWEEN CEPHALOTAXINE AND ITS CONGENERS

In addition to the alkaloids shown in Figure 1, a number of substances closely related to cephalotaxine (1) have been isolated from Cephalotaxus plants[1,4,19]. These include cephalotaxinone (11), desmethylcephalotaxinone (12), and desmethylcephalotaxine (13). The potential metabolic relationships between these four alkaloids are summarized in Figure 11. An investigation of the interrelationships has been carried out in the hope of clarifying some of the factors which regulate the concentration of cephalotaxine in Cephalotaxus plants.

[8-^{14}C]-DL-Cephalotaxine, [8-^{14}C]-DL-cephalotaxinone, and [8-^{14}C]-DL-desmethylcephalotaxinone were synthesized with the aid of the Weinreb cephalotaxine synthesis (Figure 12)[20]. Difficulty was encountered in our initial attempts to prepare [8-^{14}C]-desmethylcephalotaxine. Acid-catalyzed hydrolysis of unlabeled cephalotaxine according to a published procedure[1] yielded only desmethylcephalotaxinone (Figure 13). When the hydrolysis was carried out under argon to avoid air oxidation, the product of the reaction proved to be an inseparable mixture whose NMR spectrum indicated the presence of four compounds: the two C-7 epimers

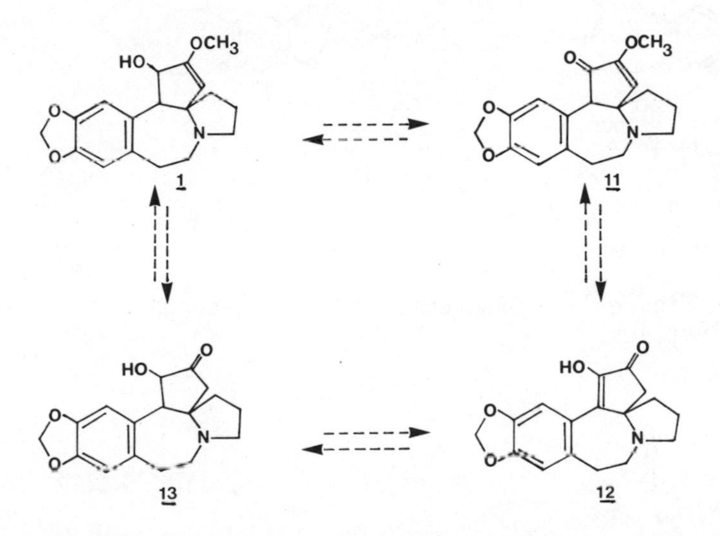

Figure 11. Potential interrelationships between cephalotaxine and its congeners.

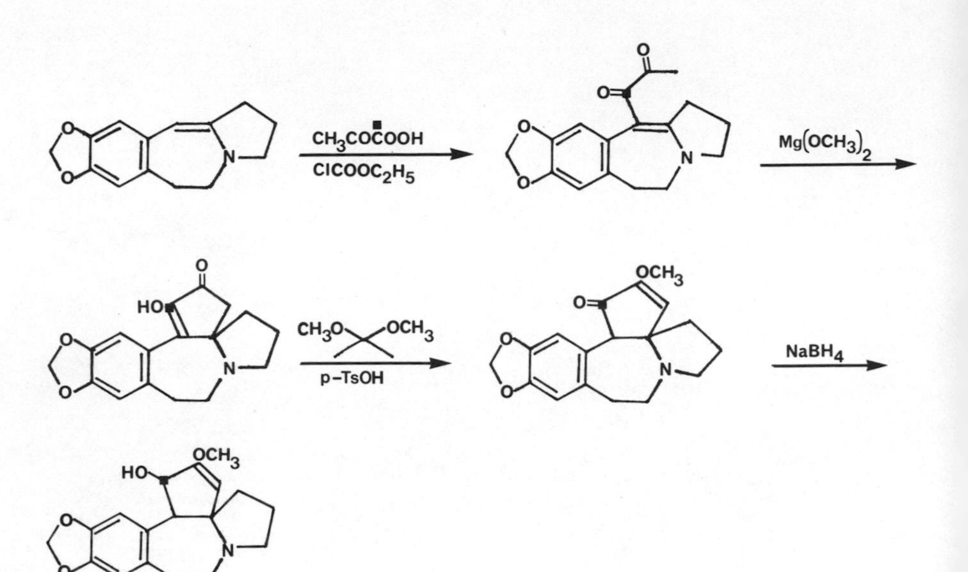

Figure 12. Synthesis of [8-^{14}C]-cephalotaxine and derivatives.

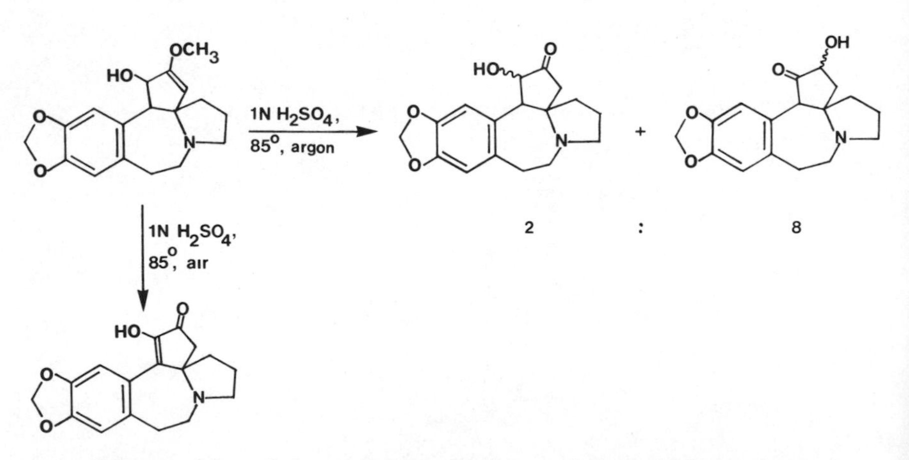

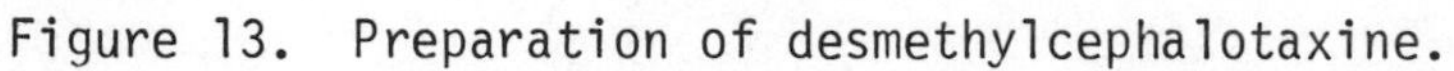
Figure 13. Preparation of desmethylcephalotaxine.

of desmethylcephalotaxine and the two C-8 epimers of the isomeric hydroxy ketone (Figure 13). The ratio between the epimeric mixture of desmethylcephalotaxines and the isomeric hydroxy ketones was approximately 2:8.* Repetition of the hydrolysis using [8-^{14}C]-DL-cephalotaxine provided a mixture containing [8-^{14}C]-DL-desmethylcephalotaxine.

A series of precursor incorporation experiments were carried out utilizing the labeled alkaloids obtained in the manner just described. The results of these experiments are summarized in Table 2. The radioactive alkaloids isolated in each experiment were degraded by conversion to the cyclopentanone 10 using known reactions (Figures 4, 12)[12,20], addition of phenyllithium, and oxidation of the adduct to yield benzoic acid. The radioactive desmethylcephalotaxine isolated in experiments 4 and 6 (Table 2) was first oxidized to desmethylcephalotaxinone by heating in dilute acid in air and the desmethylcephalotaxinone was then degraded in the usual manner. The degradations proved that the observed incorporations were specific in each case.

Experiments 1 and 2 in Table 2 show that cephalotaxine and cephalotaxinone are interconvertible in *Cephalotaxus* and that demethylation of cephalotaxinone to desmethylcephalotaxinone also takes place readily. Experiment 3 indicates that *Cephalotaxus* plants will not methylate desmethylcephalotaxinone to yield cephalotaxinone or cephalotaxine. Similarly, experiments 4 and 5 show that *Cephalotaxus* plants are capable of demethylating cephalotaxine to yield desmethylcephalotaxine, whereas they are apparently incapable of carrying out the reverse reaction, the methylation of desmethylcephalotaxine to cephalotaxine. Finally, experiment 6 demonstrates that desmethylcephalotaxinone is readily reduced by *Cephalotaxus* to desmethylcephalotaxine. The reverse reaction, oxidation of desmethylcephalotaxine to desmethylcephalotaxinone, was not examined owing to the difficulties associated with distinguishing between an *in vitro* air oxidation and an *in vivo* biological oxidation.

* The preponderance of the isomeric hydroxy ketone mixture over the mixture of desmethylcephalotaxines presumably indicates that the former compounds are more stable than the latter. The reason for this difference in stability is unclear from an examination of models.

Table 2. Administration of Labeled Alkaloids to *C. harringtonia* var. *fastigiata*.

Expt. No.	Alkaloid Administered	Feeding Period (days)	Alkaloids Isolated (% incorpor.)	% of Label at C-8
1	[8-^{14}C]-DL-Cephalotaxine	4	Cephalotaxinone (0.44)	95
			Desmethylcephalotaxinone (0.045)	99
2	[8-^{14}C]-DL-Cephalotaxinone	4	Cephalotaxine (28)	99
			Desmethylcephalotaxinone (4.6)	102
3	[8-^{14}C]-DL-Desmethyl-cephalotaxinone	4	Cephalotaxine (0)	-
			Cephalotaxinone (0)	
4	[8-^{14}C]-DL-Cephalotaxine	4	Desmethylcephalotaxine (0.79)	100
5	[8-^{14}C]-DL-Desmethyl-cephalotaxine	4	Cephalotaxine (0)	-
6	[8-^{14}C]-DL-Desmethyl-	4	Desmethylcephalotaxine (13)	98

The following conclusions can be drawn from the experimental results just discussed. First, it appears that desmethylcephalotaxine is a natural product which occurs in *Cephalotaxus* plants. Previous evidence for the natural occurrence of desmethylcephalotaxine appears tenuous in view of the fact that the material isolated from *Cephalotaxus* was identified by comparison with a sample produced by acid-catalyzed hydrolysis in the presence of air[1,4]. Second, the data suggest that neither desmethylcephalotaxine nor desmethylcephalotaxinone can be intermediates on the pathway to cephalotaxine. Rather, the enol ether methyl group present in cephalotaxine is probably introduced at an earlier stage in the biosynthesis. Finally, it appears that desmethylcephalotaxine and desmethylcephalotaxinone represent catabolites of cephalotaxine. The irreversible formation of these compounds by demethylation is analogous to the situation which obtains in the final stages of the biosynthesis of the morphine alkaloids. In the case of the latter alkaloids, it has been shown that codeine and morphine are formed by successive *O*-demethylations of thebaine[21-23].

BIOSYNTHESIS OF THE ACYL PORTIONS OF DEOXYHARRINGTONINE, ISOHARRINGTONINE, AND HARRINGTONINE

The key to the elucidation of the biosynthesis of the acyl portions of the antileukemic *Cephalotaxus* alkaloids was provided by the recognition that the diacid **18** linked to cephalotaxine in deoxyharringtonine, henceforth referred to as deoxyharringtonic acid, bears a close resemblance to an intermediate involved in the biosynthesis of leucine from valine in micro-organisms[24-29]. On the basis of this resemblance, we arrived at the hypothesis for deoxyharringtonic acid biosynthesis shown in Figure 14. Three of the putative intermediates in this hypothetical pathway were evaluated[30]. A sample of the racemic diacid **14** (Figure 14) was synthesized from 4-methyl-2-pentanone using methods developed for the synthesis of deoxyharringtonic acid from 5-methyl-2-hexanone[3]. [1-^{14}C]-*L*-Leucine was administered to *Cephalotaxus* plants and the plants harvested after 14 days. The plant material was extracted with ethanol and radioinactive **14** added as carrier. The crude mixture from the ethanol extract was saponified to convert any esters of **14** into

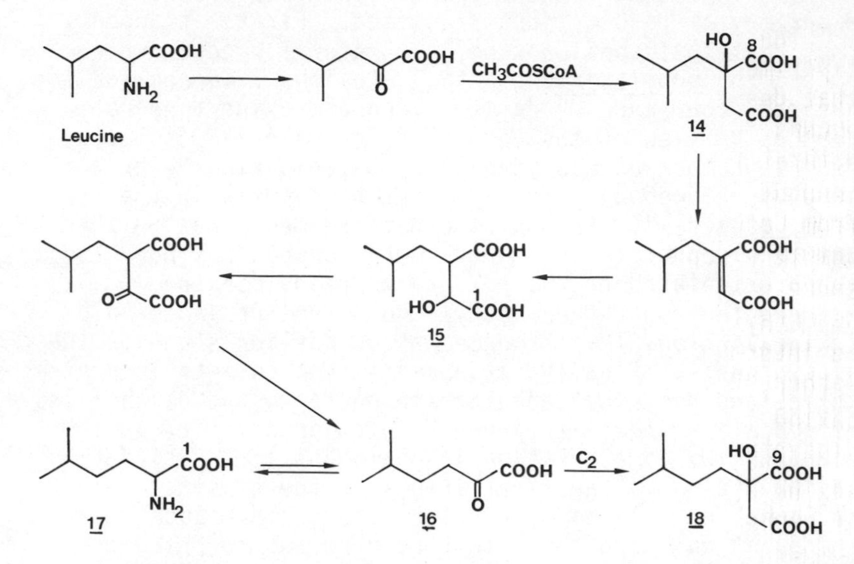

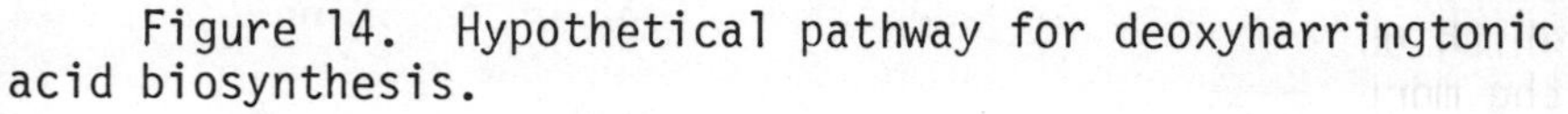
Figure 14. Hypothetical pathway for deoxyharringtonic acid biosynthesis.

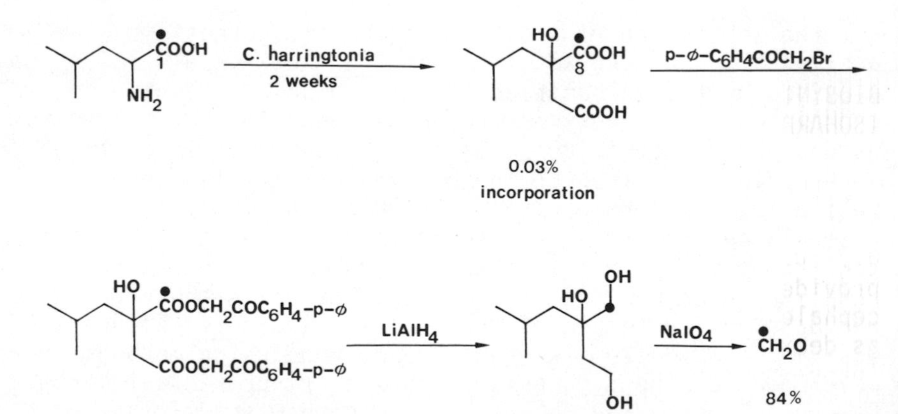

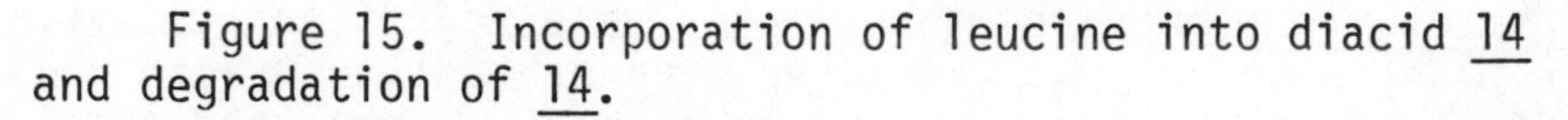
Figure 15. Incorporation of leucine into diacid 14 and degradation of 14.

the free diacid, and the crude mixture of organic acids obtained was then derivatized with p-phenylphenacyl bromide (Figure 15). The bis(p-phenylphenacyl) ester was extensively purified to give the incorporation figure listed in Table 3, expt. 1. The incorporation of leucine into diacid 14 was shown to be specific by degradation of the derivative to isolate C-8 of the diacid (Figure 15). Reduction of the bis(p-phenylphenacyl) ester with lithium aluminum hydride gave a triol which was purified by chromatography and distillation and then cleaved with periodate to yield radioactive formaldehyde, trapped as its dimedone derivative. The specific activity of the dimedone-formaldehyde showed that most of the radioactivity resided in the expected position of diacid 14 (Table 3, experiment 1). Therefore, it appears that the diacid 14 is present in Cephalotaxus plants and that it is formed from leucine in the manner anticipated.

The hypothesis delineated in Figure 14 also predicts that the diacid 15 should lie on the pathway to deoxyharringtonic acid (18). In order to test this prediction, a synthesis of [1-^{14}C]-15 was devised (Figure 16). The product of this synthesis was a mixture of diastereomers that was adequate for our purposes. Administration of [1-^{14}C]-14 to Cephalotaxus plants was followed after 7 days by ethanol extraction and addition of radioinactive deoxyharringtonic acid as carrier. After saponification, the crude mixture of acids was converted to the methyl esters and dimethyl deoxyharringtonate isolated by chromatography. The purified dimethyl ester was saponified to yield the free diacid 18 which was then converted to the bis(p-bromophenacyl) ester. Purification and degradation (Figure 17) of the bis(p-bromophenacyl) ester yielded the results presented in Table 3, experiment 2. These results clearly support the intermediacy of diacid 15 in the biosynthesis of deoxyharringtonic acid.

On the basis of Figure 14, the α-keto acid 16 should be the immediate precursor of deoxyharringtonic acid. If this were the case, then homoleucine (17) should be specifically incorporated into deoxyharringtonic acid owing to the facile interconversion between α-amino acids and the corresponding α-keto acids. [1-^{14}C]-DL-Homoleucine was prepared from 4-methylpentanal by means of a Bucherer synthesis with potassium [^{14}C-] cyanide. Administration of the labeled homo-

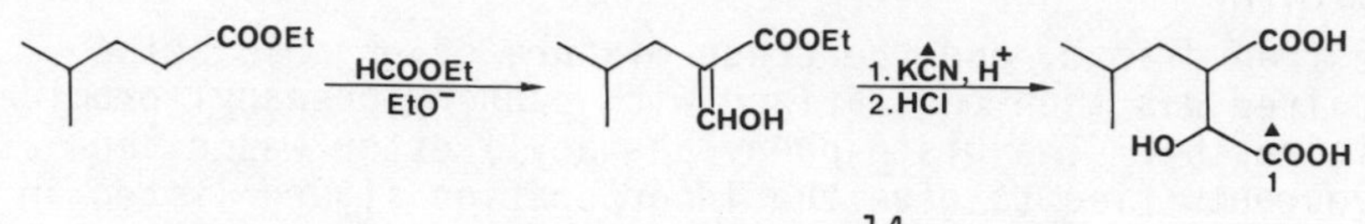

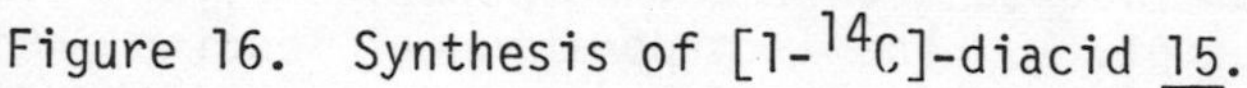

Figure 16. Synthesis of [1-^{14}C]-diacid 15.

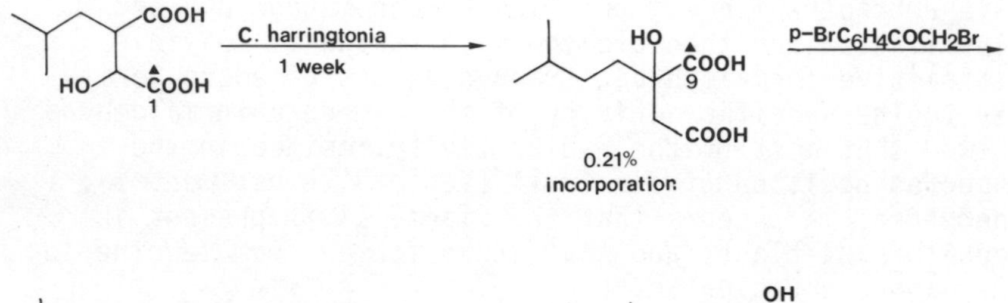

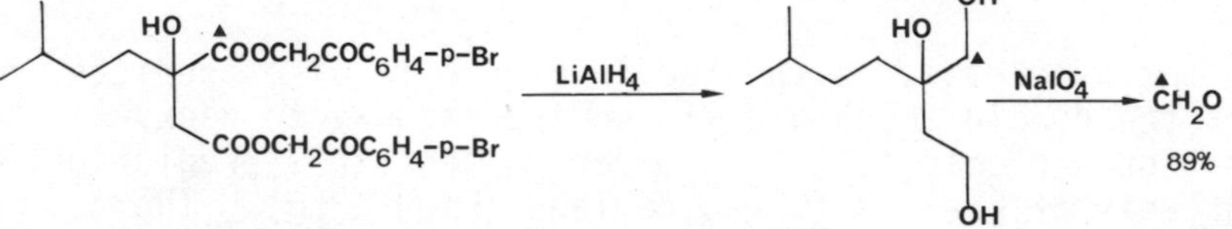

Figure 17. Incorporation of diacid 15 into deoxyharringtonic acid (18) and degradation.

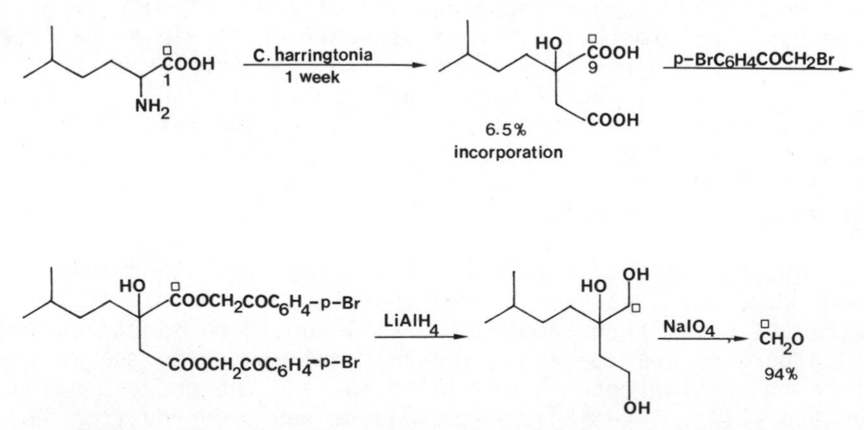

Figure 18. Incorporation of homoleucine into 18 and degradation.

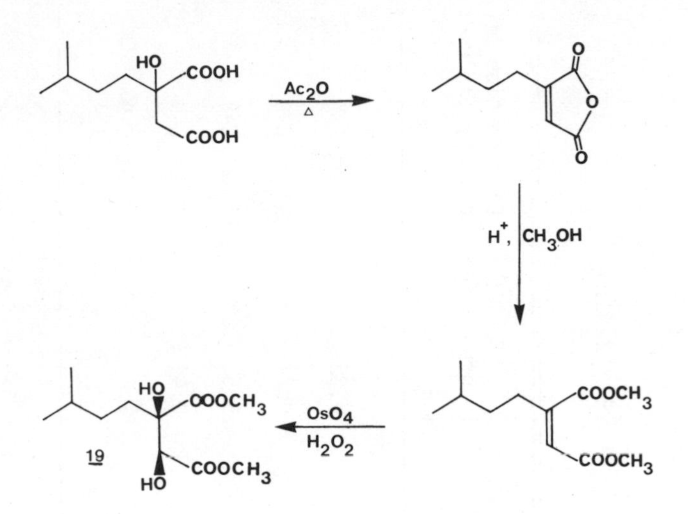

Figure 19. Synthesis of dimethyl isoharringtonate.

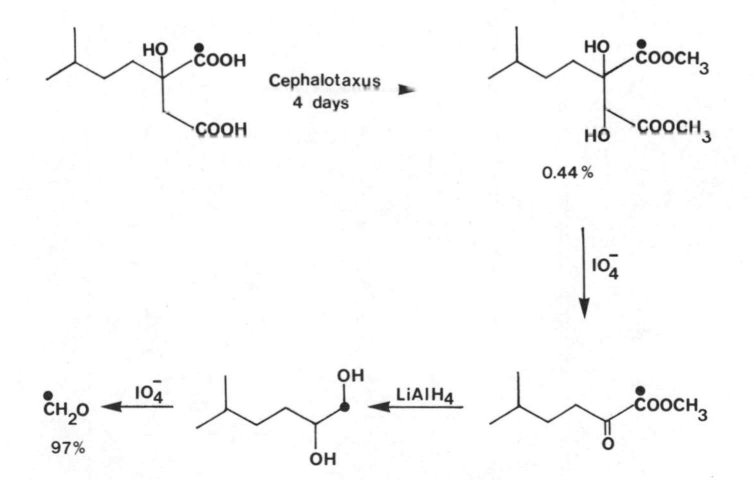

Figure 20. Incorporation of deoxyharringtonic acid into isoharringtonic acid and degradation.

Table 3. Feeding Experiments with *C. harringtonia* var. *fastigiata*.

Expt. No.	Precursor Fed	Product Isolated	Feeding Period (days)	% Incorporation	Distribution of Activity on Prod.
1	[1-^{14}C]-L-Leucine	14	14	0.03	84% at C-8
2	[1-^{14}C]-15	18	7	0.21	89% at C-9
3	[1-^{14}C]-DL-17	18	7	6.5	94% at C-9
4	[9-^{14}C]-DL-18	19	4	0.44	97% at C-9
5	[9-^{14}C]-DL-18	20	4	1.6	97% at C-9

leucine to Cephalotaxus was followed by workup after 7 days in the usual way to afford radioactive deoxyharringtonic acid as the bis(p-bromophenacyl) ester (Figure 18). After extensive purification, and degradation in the usual fashion, results were obtained (Table 3, expt. 3) demonstrating that homoleucine is an efficient and specific precursor of deoxyharringtonic acid. This observation, in conjunction with the experiments evaluating diacids 14 and 15, provides convincing evidence that deoxyharringtonic acid (18) is biosynthesized according to the pathway outlined in Figure 12.

The diacids which are linked to cephalotaxine in isoharringtonine (4) and harringtonine (2) (Figure 1) appear to be derived from deoxyharringtonic acid by hydroxylation. Evidence supporting this possibility was obtained in the following way. [9-^{14}C]-DL-Deoxyharringtonic acid was synthesized from 1-carbethoxy-5-methyl-2-hexanone[3] by treatment with potassium [^{14}C-] cyanide and acid-catalyzed hydrolysis of the intermediate cyanohydrin. Radioinactive isoharringtonic acid dimethyl ester (19) was synthesized from deoxyharringtonic acid by the route summarized in Figure 19. Administration of [9-^{14}C]-DL-deoxyharringtonic acid to Cephalotaxus for 4 days was followed by workup with addition of unlabeled dimethyl isoharringtonate as carrier. The crude plant extract was saponified and the acidic fraction reesterified with diazomethane. Dimethyl isoharringtonate was isolated by chromatography and purified by repeated recrystallization. Degradation of the labeled dimethyl ester obtained in this fashion yielded radioactive formaldehyde (Figure 20) which was trapped as its dimedone derivative. The results of this incorporation experiment (Table 3, expt. 4) demonstrate that isoharringtonic acid is derived in vivo from deoxyharringtonic acid, presumably by direct hydroxylation.

Harringtonic acid was prepared by modification of a synthesis[31] devised Kelly (Figure 21). Free harringtonic acid exists as the δ-lactone 20, which is not a readily crystallizable substance. However, treatment of the lactone with strong acid yields the ether diacid 21, which is beautifully crystalline. [9-^{14}C]-DL-Deoxyharringtonic acid was fed to Cephalotaxus plants and radioinactive δ-lactone 20 added during workup. The crude extract was subjected to acidic saponification and radioactive ether diacid 21

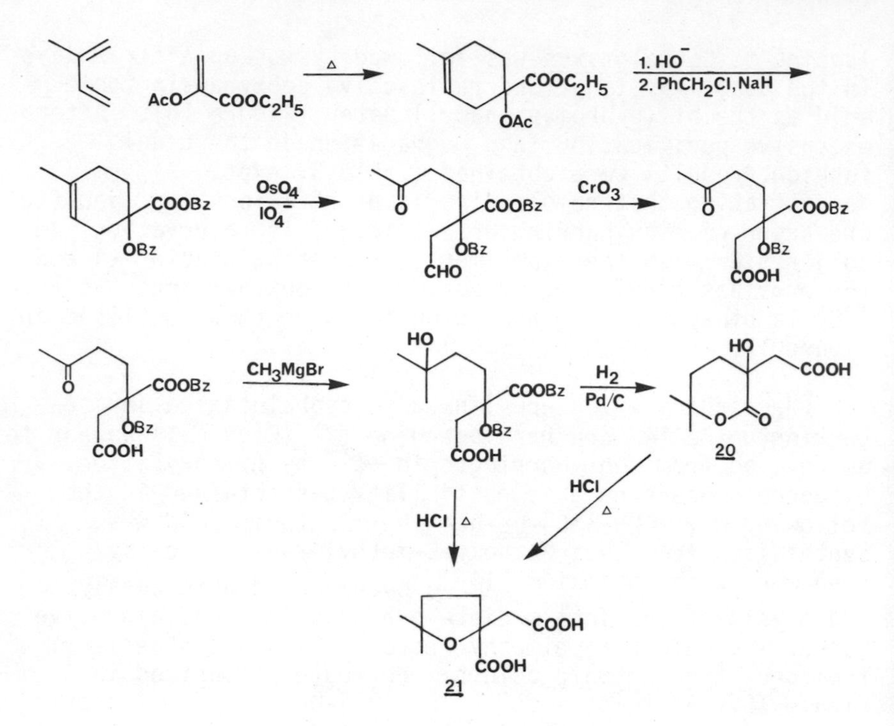

Figure 21. Synthesis of harringtonic acid.

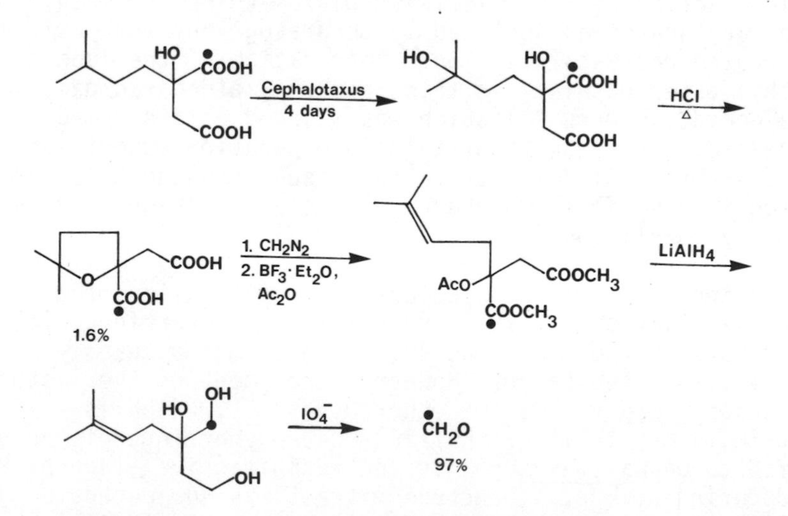

Figure 22. Incorporation of deoxyharringtonic acid into harringtonic acid and degradation.

isolated by chromatography. The ether diacid was purified extensively and then degraded by the methods portrayed in Figure 22 to give the results listed in Table 3, expt. 5. The data clearly show that harringtonic acid is derived *in vivo* from deoxyharringtonic acid, probably by hydroxylation.

CONCLUDING REMARKS

The results of the experiments discussed in this paper represent the current state of knowledge with regard to the biosynthesis of the *Cephalotaxus* alkaloids. A number of problems remain to be investigated before this knowledge will be in any way complete. Nothing is known at present, for example, concerning the stages involved in the formation of the antileukemic ester alkaloids, and even greater ignorance surrounds the problem posed by the nature of the intermediate stages in cephalotaxine biosynthesis. It is to be hoped that future investigations will lead to clarification of these remaining areas of uncertainty.

ACKNOWLEDGEMENTS

No account of our work would be complete without the grateful acknowledgement of the efforts of John Schwab, Michael Chang, Amy Gitterman, Daniel Sternbach, Richard Dufresne, and Michael Cabelli, all of whom contributed to the results described. Thanks are also due to the NIH for generous financial support (GM-19220) and a Career Development Award (GM-00143), to the NSF for funds used to purchase the Bruker NMR spectrometer used in much of this work (grants GU3852, GP37156), to Professor S. Weinreb for a gift of the intermediate used to prepare labeled cephalotaxine derivatives, to Dr. R. G. Powell for gifts of alkaloids, and to Dr. Robert Perdue for *Cephalotaxus* plants.

REFERENCES

1. Paudler, W. W., G. I. Kerley, and J. McKay. 1963. The Alkaloids of Cephalotaxus drupacea and Cephalotaxus fortunei. J. Org. Chem. 28:2194.
2. Abraham, D. J., R. D. Rosenstein, and E. L. McGandy. 1969. Single Crystal X-ray Structures of Chemotherapeutic Agents. II. The Structure of Cephalotaxine Methiodide. Tetrahedron Lett. 4085.
3. Mikolajczak, K. L., R. G. Powell, and C. R. Smith, Jr. 1972. Deoxyharringtonine, a New Antitumor Alkaloid from Cephalotaxus: Structure and Synthetic Studies. Tetrahedron 28:1995.
4. Paudler, W. W., and J. McKay. 1973. The Structures of some of the Minor Alkaloids of Cephalotaxus fortunei. J. Org. Chem. 38:2110.
5. Powell, R. G., D. Weisleder, C. R. Smith, Jr., and W. K. Rohwedder. 1970. Structures of Harringtonine, Isoharringtonine, and Homoharringtonine. Tetrahedron Lett. 815.
6. Powell, R. G., D. Weisleder, and C. R. Smith, Jr. 1972. Antitumor Alkaloids from Cephalotaxus harringtonia: Structure and Activity. J. Pharm. Sci. 61:1227.
7. Powell, R. G. 1972. Structures of Homoerythrina Alkaloids from Cephalotaxus harringtonia. Phytochemistry 11:1467.
8. Johns, S. R., J. A. Lamberton, and A. A. Sioumis. 1969. Alkaloids of Schelhammera Pedunculata (Liliaceae). III. The Structures of Schelhammericine and Alkaloids A, B, E, G, H, J, and K. Aust. J. Chem. 22:2219.
9. Barton, D. H. R., R. B. Boar, and D. A. Widdowson. 1970. Phenol Oxidation and Biosynthesis, Pt. XXI. The Biosynthesis of Erythrina Alkaloids. J. Chem. Soc. C:1213.
10. Barton, D. H. R., R. James, G. W. Kirby, D. W. Turner, and D. A. Widowson. 1968. Phenol Oxidation and Biosynthesis, Pt. XVII. The Structure and Biosynthesis of Erythrina Alkaloids. J. Chem. Soc. C:1529.
11. Battersby, A. R., R. B. Herbert, E. McDonald, R. Ramage, and J. H. Clements. 1972. Alkaloid Biosynthesis, Pt. XVIII. Biosynthesis of Colchicine from the 1-Phenethylisoquinoline System. J. Chem. Soc., Perkin I. 1741.

12. Parry, R. J., and J. M. Schwab. 1975. Biosynthesis of Cephalotasus Alkaloids. I. Novel Mode of Incorporation of Tyrosine into Cephalotaxine. J. Am. Chem. Soc. 97:2555.
13. Schwab, J. M., M. N. T. Chang, and R. J. Parry. 1977. Biosynthesis of Cephalotaxus Alkaloids. 3. Specific Incorporation of Phenylalanine into Cephalotaxine. J. Am. Chem. Soc. 99:2368.
14. Battersby, A. R., E. McDonald, J. A. Milner, S. R. Johns, J. A. Lamberton, and A. A. Sioumis. 1975. Biosynthesis of Schelhammeridine: Mode of Specific Incorporation of [2-^{14}C]-Tyrosine. Tetrahedron Lett. 3419.
15. Wightman, R. H., J. Staunton, A. R. Battersby, and K. R. Hanson. 1972. Studies of Enzyme-mediated Reactions, Pt. 1. Synthesis of Deuterium-or Tritium-Labeled (3S)- and (3R)-Phenylalanines: Stereochemical Course of the Elimination Catalyzed by L-Phenylalanine Ammonia Lyase. J. Chem. Soc., Perkin Trans. I:2355.
16. Retey, J., K. Bartl, E. Ripp, and W. E. Hull. 1977. Stereospecificity of Phenylpyruvate Tautomerase. Eur. J. Biochem. 72:251.
17. Hales, N. J., and H. Heaney. 1975. The Preparation and Assay of Specifically [^{14}C]-labeled Benzene Derivatives. Tetrahedron Lett. 4075.
18. Kratzl, K., and F. W. Vierhapper. 1971. Spezifisch ^{14}C-Kernmarkierte Phenolderivate, 1. Mitt.: Synthese von [^{14}C]-Guajacol. Monat. Chem. 102:224.
19. Powell, R. G., and K. L. Milolajczak. 1973. Desmethylcephalotaxinone and Its Correlation with Cephalotaxine. Phytochemistry 12:2987.
20. Auerbach, J., and S. M. Weinreb. 1972. The Total Synthesis of Cephalotaxine. J. Am. Chem. Soc. 94:7172.
21. Rapoport, H., F. R. Stermitz, and D. R. Baker. 1960. The Biosynthesis of Opium Alkaloids. I. The Interrelationships Among Morphine, Codeine, and Thebaine. J. Am. Chem. Soc. 82:2765.
22. Parker, H. I., G. Blaschke, and H. Rakpoport. 1972. The Biosynthesis of Opium Alkaloids. I. The Interrelationships among Morphine, Codeine, and Thebaine. J. Am. Chem. Soc. 94:1276.
23. Stermitz, F. R., and H. Rakpoport. 1961. The Biosynthesis of Opium Alkaloids. Alkaloid Intercon-

versions in *Papaver somniferum* and *Papaver orientale*. *J. Am. Chem. Soc.* *83*:4045.
24. Calvo, J. M., M. G. Kalyanpur, and C. M. Stevens. 1962. 2-Isopropylmalate and 3-Isopropylmalate as Intermediates in Leucine Biosynthesis. *Biochemistry* *1*:1157.
25. Calvo, J. M., C. M. Stevens, M. G. Kalyanpur, and H. E. Umbarger. 1964. The Absolute Configuration of α-Hydroxy-β-carboxyisocaproic Acid (3-Isopropylmalic Acid), an Intermediate in Leucine Biosynthesis. *Biochemistry* *3*:2024.
26. Gross, S. R., R. O. Burns, and H. E. Umbarger. 1963. The Biosynthesis of Leucine. II. The Enzymatic Isomerization of β-Carboxy-β-hydroxy-isocaproate and α-Hydroxy-β-carboxyisocaproate. *Biochemistry* *2*:1046.
27. Gross, S. R., C. Jungwirth, and H. E. Umbarger. 1962. Another Intermediate in Leucine Biosynthesis. *Biochem. Biophys. Res. Commun.* *7*:5.
28. Jungwirth, C., P. Margolin, and H. E. Umbarger. 1961. The Initial Step in Leucine Biosynthesis. *Biochem. Biophys. Res. Commun.* *5*:435.
29. Strassman, M., and L. N. Ceci. 1963. Enzymatic Formation of α-Isopropylmalic Acid, an Intermediate in Leucine Biosynthesis. *J. Biol. Chem.* *238*:2445.
30. Parry, R. J., D. D. Sternbach, and M. D. Cabelli. 1976. Biosynthesis of Cephalotaxus Alkaloids. 2. Biosynthesis of the Acyl Portion of Deoxyharringtonine. *J. Am. Chem. Soc.* *98*:6380.
31. Kelly, T. R., R. W. McNutt, Jr., M. Montury, N. P. Tosches, K. L. Mikolajczak, C. R. Smith, Jr., and D. Weisleder. 1978. The Preparation of Harringtonine from Cephalotaxine. Submitted for publication.

Chapter Four

ENZYMOLOGY OF INDOLE ALKALOID BIOSYNTHESIS

K. M. MADYASTHA AND CARMINE J. COSCIA

E. A. Doisy Department of Biochemistry
Saint Louis University School of Medicine
St. Louis, Missouri 63104

INTRODUCTION

The monoterpenoid indole alkaloids are a large family of structurally diverse compounds encompassing some of the most important plant medicinals discovered by man. Although these alkaloids were one of the last major groups of well-known natural products to have their biogenesis delineated, they have become one of the first classes of plant secondary metabolites to be studied at the enzymatic level. A contributory factor has been the excellent progress made by *in vivo* tracer work, which within a little more than a decade has led to the elucidation of the major features of a highly complicated pathway[1,2].

In *Catharanthus roseus* (L.) G. Don (also known as *Vinca rosea*) the metabolic scenario entails fabrication of four different classes of indole alkaloids and their dimers (including vinblastine and vinchristine), from the amino acid tryptophan, a monoterpene and the methyl group of methionine (Figure 1). Our contribution to this body of knowledge in terms of *in vivo* tracer studies has focused on stereospecific aspects of the pathway from mevalonate to the monoterpene glucosides[3], and our enzyme work has also been centered on these early stages. This chapter contains a discussion of a cytochrome *P*-450-dependent monoterpene hydroxylase which converts the acyclic alcohols, geraniol and nerol, to their corresponding 10-hydroxy derivatives, which have been demonstrated to be precursors of loganin and indole alkaloids in *C. roseus* by *in vivo* tracer studies[4,5]. Investigations of a pyridine nucleotide-dependent monoterpene alcohol oxidoreductase and *S*-adenosyl-*L*-methionine; loganic acid methyl transferase, a carboxyl-methylating enzyme, will also be reviewed. In recent studies in other laboratories the enzymological approach has been extended beyond this point to examine the synthesis of the first alkaloid formed in the pathway, strictosidine, and its conversion to ajmalicine and epimers (Figure 1). Tryptamine and secologanin condense to form the β-carboline, strictosidine, which is metabolized to cathenamine by a β-glucosidase and the latter reduced in a pyridine nucleotide-dependent reaction to the *Corynanthe* alkaloids.

Despite the fact that urease, the first enzyme to be crystallized, was isolated from the jack bean by Sumner in 1926[6], plant enzymology has not kept pace with the rapid growth of mammalian or bacterial enzyme purification and characterization. This is particularly true of membrane-bound enzymes and those catalyzing secondary metabolism[7]. Numerous reasons may be cited to account for this. They include lack of sufficient amounts of plant material, substrates, and products, the awesome number of metabolic transformations involved, the difficulty in rupturing cellulose-containing cell walls with concomitant retention of intact organelles such as fragile vacuoles, and the presence of large amounts of secondary metabolites which together with phenoloxidases[8] and other degradative enzymes can rapidly inactivate other enzymes. On the other hand, progress in enzymology has been such that enzyme isolation and purification to homogeneity is no longer as formidable

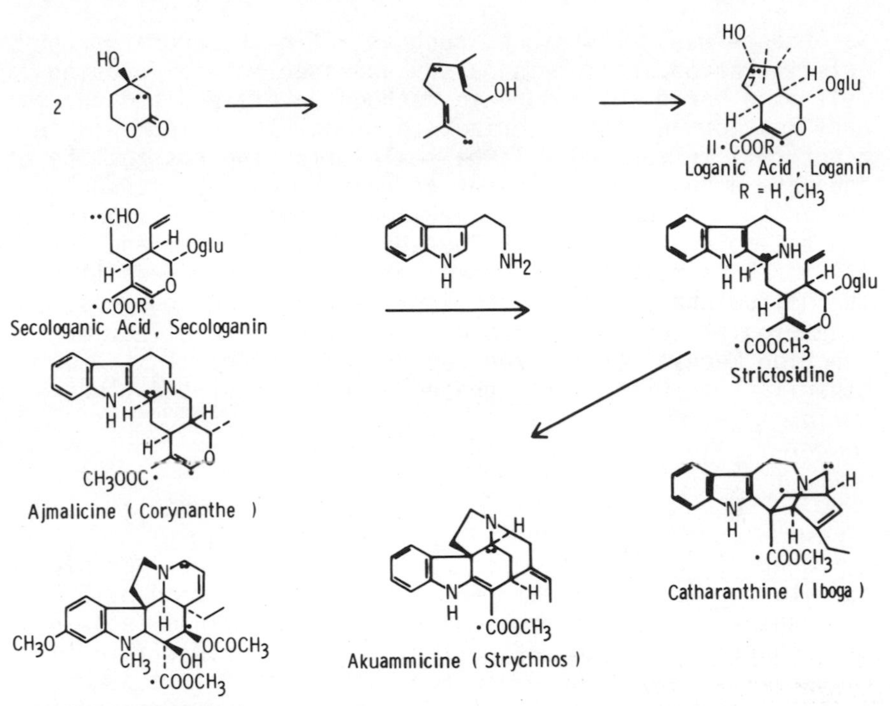

Figure 1. Synopsis of the Biosynthesis of Indole Alkaloids.

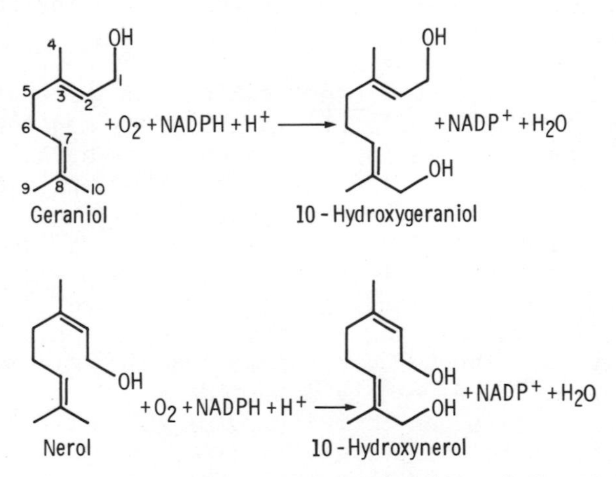

Figure 2. Enzymatic Hydroxylation of Monoterpene Alcohols.

as it once was. Techniques such as affinity chromatography, gel filtration, radioimmunoassay and isoelectric focusing as well as other electrophoretic methods have expedited enzyme purification and characterization. Recent developments in plant cell culture on a large scale offer the possibility of adequate amounts of tissue as an enzyme source. It would appear then to be an opportune time to inaugurate studies of the enzymology of secondary metabolism in plants, and efforts by a number of groups are under way. In addition to the unique challenges of the isolation of plant enzymes, the enzymologist can look forward to uncovering novel enzyme reaction mechanisms and regulation mediated by light and other factors with little precedent outside of the plant world.

MONOTERPENE HYDROXYLASE

General Considerations. The hydroxylation of geraniol and its *cis* isomer, nerol, (Figure 2) is catalyzed by a monoxygenase isolated from three to eight-day-old seedlings of *Catharanthus roseus* as well as mature plants (Table 1). Five-day-old seedlings exhibited maximal activities, whereas by the eighth day, values had dropped to those observed in mature plants. Monoterpene hydroxylase activity was detected in plants that synthesize other indole alkaloids (*Vinca minor*) and monoterpene glucosides (*Lonicera morrowi*) but not in *Pisum sativum*, a plant which makes neither of these secondary metabolites.

Preparations from *C. roseus* callus and suspension cultures were also capable of hydroxylating monoterpenes but at a lower rate. Isolation of the membrane-bound oxygenase from *C. roseus* tissue culture sources was complicated by an apparent change in the physical properties of cell walls and membranes. Thus satisfactory disruption of callus or cell suspensions was only achieved with a French press and the various membrane fractions in the resultant homogenates did not always sediment in the expected manner upon differential centrifugation. The soluble enzyme loganic acid methyltransferase was more readily isolated from *C. roseus* callus tissue and exhibited an activity comparable to that of similarly treated seedling tissue (13.5 nmol/h/mg protein)[12]. In a few instances hydroxylase activities three times as high were obtained from the usual 20,000 *g* pellets from *C.*

Table 1. Survey of Various Plant Tissues for Monoterpene Hydroxylase and NADPH-Cytochrome c Reductase Activity

Source	Hydroxylase Activity* (nmol/min/mg prot.)	Reductase Activity (nmol/min/mg prot.)
Catharanthus roseus (5-day-old seedlings)	0.50	18.4
Catharanthus roseus (mature plants)	0.16	23.0
Catharanthus roseus (callus tissue from seedlings)**	0.06	14.5
Catharanthus roseus (suspension cultures from callus)**	0.06	25.0
Vinca minor (mature plants)	0.03	----
Lonicera morrowi (mature plants)	0.02	----
Pisum sativum (seedlings)	0	----

*A 20,000 g pellet was prepared from tissue extracted with 0.1 M Tris•HCl, pH 7.6, containing 0.4 M sucrose, 1 mM DTT, 10 mM KCl, 10 mM $Na_2S_2O_5$, 10 mM $MgCl_2$ and 10 mM EDTA[9]. Values given were the highest obtained.

**Cultures were initiated from intact seedlings, excised roots or cotyledons[10, 11].

roseus cell suspension cultures, suggesting that under appropriate conditions it may be possible to achieve better recoveries. Indeed much remains to be learned about this overall approach. For example, most C. roseus tissue culture lines developed thus far have been reported to yield identifiable alkaloids of the Corynanthe family alone[11,13,14]. NADPH-cytochrome c reductase activities were maximal in cell suspensions (Table 1). Several common enzymes are capable of transferring electrons to cytochrome c including the reductase component of P-450 heme protein-dependent mixed-function oxidases. It would be of interest to determine what percentage of the callus activity is attributable to the presence of a cytochrome P-450 reductase activity.

The allylic hydroxylation of geraniol and nerol by membranous fractions of C. roseus required O_2 and NADPH and was sensitive to typical inhibitors of that group of monoxygenases which utilize cytochrome P-450 as a terminal oxidase[9,15]. Most diagnostic was the inhibition by CO, which could be reversed selectively by light having a maximal intensity between 420 and 450 nm. One of the first plant enzymes demonstrated to be cytochrome P-450-dependent was kaurene hydroxylase[16,17]. Plant N-demethylase[18,19], cinnamic acid-4-hydroxylase[20-23], and a fatty acid epoxidase[24] and hydroxylase[25] also appear to belong to this class of mixed function oxidases. Cytochrome P-450 content in plants seems to vary markedly not only between species but also in different organs within the same species[26,27]. This is consistent with the existence of a heterogeneous population of P-450 heme proteins with different substrate specificities participating in both primary and secondary metabolism[28]. It would also parallel the apparent diversity in plants of cytochrome b_5-type proteins. Many of the reactions catalyzed by P-450 systems in mammals, e.g., O-demethylation and fatty acid and steroid hydroxylations, also occur in plants. The existence of corresponding P-450-dependent enzymes in plants would be consistent with the extensive evolutionary conservatism in nature. Studies such as that of the liver microsomal monoterpene hydroxylase and the C. roseus NADPH-cytochrome c (P-450) reductase described below are also supportive of this viewpoint.

Despite the high sensitivity of the CO binding assay used to detect the pigment (our lower limit of detection:

ΔOD 450-490 = 6 pmoles) little or no P-450 heme protein has been found in many plant microsomal preparations screened so far[27]. Various hypotheses can be invoked to explain these observations but speculation should be reserved until more knowledge accrues on plant membrane preparations. Since P-450 heme proteins are labile and very susceptible to the action of hydrolases, especially proteases, there is the possibility of their rapid degradation in cell-free homogenates. In comparative studies in our laboratories, hepatic microsomal cytochrome P-450 has been found to be far more stable than its plant counterpart. The lability of the latter is enhanced upon solubilization and fractionation by DEAE cellulose (see below). In fact, under certain conditions the P-450 heme protein is not denatured to the frequently observed catalytically-inactive P-420 form[26] but a total loss of the Soret band in difference spectra measured in the presence of reducing equivalents and CO occurs, suggesting that the heme prosthetic group is removed from the apoenzyme. Such lability has been observed for $P\text{-}450_{CAM}$ present in the absence of stabilizing substrate during isolation[29]. Different conditions are often required to isolate P-450 heme proteins from various mammalian tissues. The brain is exemplary as an organ which was believed to be devoid of this cytochrome until its recent discovery in concentrations comparable to those of *C. roseus* seedlings (60-100 pmol/mg protein)[30,31].

Another factor to be considered when studying plant cytochrome P-450 is the co-occurrence of another CO-binding heme protein in microsomal membranes[32]. This heme protein, recently isolated from *Pisum sativum* seedlings, is distinguished from cytochrome P-450 by its facile reduction by H_2O_2 in the presence of CO to give a product with a difference spectrum with maxima at 425, 535, and 570 nm. Destruction of this new heme protein is accompanied by bleaching of carotenoids and it exhibits a high affinity for various amines. Circumstantial evidence exists to suggest this heme protein may also be present in *C. roseus* seedlings, tubers of *Solanum tuberosum* and spadices of *Arum italicum*. This is based on time course studies of the changes in difference spectra[33,34], during reduction in presence of CO. This time-dependent appearance of bands at 420 and 485 nm without change in the 450 nm absorbance (see Figure 2 in Ref.[9]) has been attributed to enzymic degradation of carotenoids in the reference cuvette which is inhibited by CO in the sample cuvette. Carotenoids with structures similar to lutein give maximal absorption at 420 and 485 nm and are found in plant microsomes and vacuolar membranes[9,32,35,36]. Metyrapone can replace CO in producing this effect[9]. Solubilized fractions of the C. roseus putative carotenoid degradative system have been separated from cytochrome P-450 by affinity chromatography on aminohexyl Sepharose columns[37]. Additional studies are needed to support this possible relationship and determine the physiological significance of this new heme protein.

Substrate Specificity. Original studies on the substrate specificity of the C. roseus monoterpene hydroxylase were designed to demonstrate that the diol fraction obtained by TLC and counted in the radiometric enzyme assay represents a single 10-hydroxy derivative (Figure 2)[15]. Upon isolation and estimation of the total radioactivity of the diol fraction, authentic "cold" 10-hydroxygeraniol or nerol were added. After reaction with 3,5-dinitrobenzoyl chloride, the ester derivative was recrystallized to constant specific activity. The final radioactivity found in the corresponding diol derivative accounted for almost all of the activity in the diol fraction from experiments in which geraniol was substrate and 85% of the activity found in those with nerol. The results of these experiments established that the C. roseus oxygenase exhibited a high specificity for hydroxylation at the C-10 (E) methyl group. These results also established that cis-trans isomerization does not occur under the conditions of the hydroxylase assay. Interestingly rabbit liver microsomal P-450LM$_2$ (phenobarbital induced) exhibits a similar specificity for the E methyl of geraniol and nerol[38]. Yet the interaction of many mono- and particularly cyclic sesquiterpenes with the heme protein as detected by optical difference spectra suggest they would be substrates for the liver microsomal P-450 system[38].

In vivo tracer experiments have provided evidence for the intermediacy of geranyl pyrophosphate[39] as well as geraniol and nerol[1] in the indole alkaloid pathway. The question then arises as to whether a phosphatase-mediated hydrolysis of the monoterpene pyrophosphate precedes hydroxylation. This would be likely considering the hydrophilic character of the monoterpene pyrophosphate and the hydrophobic preference of P-450 heme protein substrate binding sites. To this end, tritium-labeled geranyl and neryl pyrophosphate have been incubated with both *C. roseus* and mammalian membrane preparations. As shown in Figure 3 no hydroxylation was observed over a broad concentration range in which the corresponding alcohols are transformed[40]. These results are consistent with a pathway involving hy-drolysis of the monoterpene pyrophosphate followed by hy-droxylation. It also suggests that in mammalian systems the hydroxylation is catabolic. In fact, a pyrophosphatase with a specificity for geranyl and neryl pyrophosphates has been isolated from *Citrus sinensis*[41]. Mammalian liver and testes also contain pyrophosphohydrolases specific for acyclic isoprenoid pyrophosphates[42,43]. The pyrophosphatase step in plants may be reversible. *In vivo* tracer experiments have shown that administered labeled geraniol can be converted to triterpenes[44-46] in *Pisum sativum* and *Menyanthes trifoliata* and a geraniol kinase has been isolated from *Mentha piperita*[47]. Since no evidence has been put forth to demonstrate that geraniol is incorporated into triterpenes without prior degradation, e.g., to acetate[48,49], and since the kinase activity detected was relatively low, additional studies are necessary to establish reversibility unequivocally.

Solubilization and Purification of the Components of the Multienzyme Complex. In the last decade significant advances in the solubilization and purification of mammalian membrane-bound enzymes has been made. This success has set the stage for comparative studies with plant systems with the recognition that the latter can be fraught with complications not always encountered with mammalian tissues. Before detergent solubilization studies on *C. roseus* membrane fractions were initiated, hydroxylase assays were conducted in the presence of a series of ionic and nonionic detergents to determine their effect on enzyme activity[9]. Cholate and Renex were found to be least inhibitory and have been used in subsequent studies. Sonication followed by addition of either ionic or nonionic detergent has afforded soluble preparations in which up to 65% of the original hydroxylase was recovered. Chromatography on DEAE-cellulose columns has given good separation of fractions rich in *P*-450 heme protein from the NADPH-cytochrome *c* (*P*-450) reductase. The latter enzyme was found to be quite stable and could be stored at -70° C with little loss in activity. In contrast the *P*-450 heme protein fraction from the DEAE-cellulose column has proven highly labile. Nevertheless, these frac-

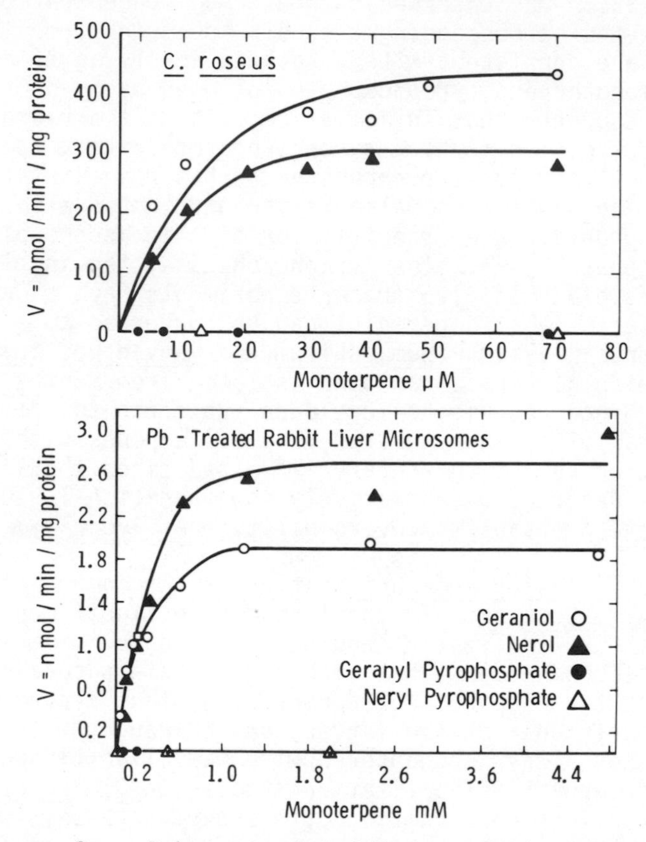

Figure 3. Substrate Specificity of the Monoterpene Hydroxylase. Incubation conditions [38,40] were modified to inhibit phosphatase activity and this was ascertained by controls. The usual TLC assay was used after treatment of the products with alkaline phosphatase and apyrase[88]. The open square represents an assay in which both geraniol and geranyl pyrophosphate were coincubated. The results indicate no change in activity, indicating competition was not exhibited by the latter.

tions have been recombined with lipid to afford active hydroxylase preparations[9].

By a sequence of calcium phosphate gel adsorption, DEAE-Sephadex A-50 fractionation and final Sephadex G-200 chromatography, a 120-fold-purified preparation of the reductase has been obtained (Table 2), which upon SDS-polyacrylamide gel electrophoresis exhibited two major polypeptide bands with molecular weights of 7.8×10^4 and 6.3×10^4 [50]. Although this reductase was found to be devoid of P-450 and b_5-type heme proteins, NADH-cytochrome c reductase, and DT diaphorase, its flavin content was very low and it did not recombine well with heme proteins in the hydroxylase assay. Addition of FMN to the reductase caused considerable stimulation of activity (Table 2). On the basis of studies with mammalian[51] and yeast reductases[52] it appeared that the plant reductase may undergo proteolytic degradation as well as loss of its flavin prosthetic group(s)[50]. Mammalian microsomal NADPH-cytochrome c reductase is an amphiphilic protein of 7.6 - 7.8×10^4 molecular weight which loses a hydrophobic region when isolated in the presence of lipase or protease[51]. The resultant NADPH-cytochrome c reductase of 6.8×10^4 molecular weight does not reduce cytochrome P-450. Thus it is possible that endogenous plant proteases were responsible for partial proteolysis of the 7.8×10^4--dalton plant species to a protein of 63,000 molecular weight. The following findings are consistent with this line of reasoning. If the plant reductase was subjected to polyacrylamide gel elecrophoresis and stained with NADPH-neotetrazolium, two pink bands were observed which corresponded to two major bands stained with Coomassie blue after electrophoresis under nondenaturing conditions. It has been reported by Fan and Masters[53] that detergent and proteolytically solubilized mammalian NADPH-cytochrome c reductase are capable of reducing neotetrazolium in the presence of NADPH. In an attempt to improve the purification and minimize the putative degradation of the P-450 reductase another approach was undertaken utilizing bioaffinity chromatography[50]. The 2'5'-ADP-Sepharose column used selectively binds hepatic microsomal NADPH-cytochrome C reductase[54]. The isolation procedure entailed solubilization and DEAE-cellulose column chromatography in the presence of Renex 690 (Figure 4) or sodium cholate followed by affinity chromatography. The reductase obtained in 36% yield was purified 745-fold and was capable of reconstituting geraniol hydroxy-

Table 2. Purification of NADPH-Cyt. c Reductase from *C. roseus*

Preparation	Total Protein (mg)	Specific Activity (nmol/min/mg)	Stimulation of Activity by Flavins (%)		Yield (%)
			FMN	FAD	
20,000 g pellet	624	12	0	0	100
Sodium cholate solubilized	301	16.6	2	0	67
DEAE-cellulose column eluate	71	119.3	7.5	6	112
Calcium phosphate gel	13	267.8	7	2	45
DEAE-Sephadex A-50 column eluate	3	586	52	10	25
Sephadex G-200 column eluate	0.6	1429	131	10	12

Enzyme assays were performed as previously described[9,50].

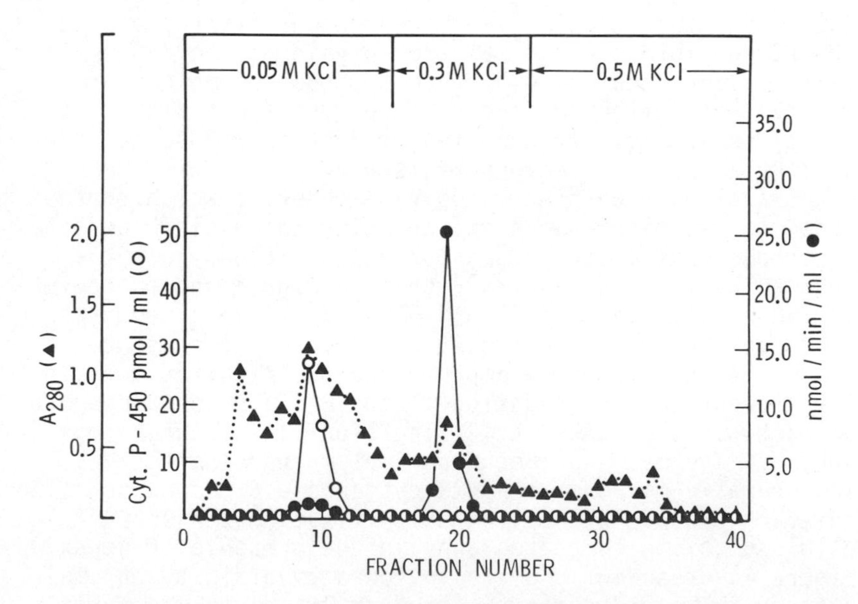

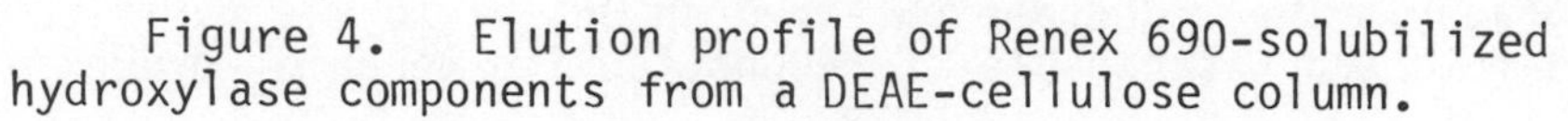
Figure 4. Elution profile of Renex 690-solubilized hydroxylase components from a DEAE-cellulose column.

Table 3. Geraniol Hydroxylation in a Reconstituted System from *C. roseus*

Fraction	Geraniol hydroxylation (nmol/min/nmol *P*-450)
Cytochrome *P*-450	0.069
Reductase	0
Lipid	0
Cytochrome *P*-450 + Reductase	0.166
Cytochrome *P*-450 + Lipid	0.104
Cytochrome *P*-450 + Reductase + Lipid	0.53

The assay mixture contained 150 μmol Tris-HCl, 5 nmol FMN, 1.5 μmol DTT, [1-^{3}H]geraniol (45 x 10^4 dpm, 11 nmol) and 0.5 μmol NADPH in a total volume of 1.5 ml. Variable fractions used were 9 pmol *P*-450, 4.3 units affinity chromatographed reductase and 0.2 mg total lipid ($CHCl_3$:MeOH, 2:1, extract). The mixture was incubated at 35° C for 30 min[50].

Table 4. Lipid Dependence of Geraniol Hydroxylation

Additive	% Activity
None (Complete)	100
Lipid (0.5 mg)	81
Phospholipase A (2 units)	26
Phospholipase C (2 units)	12
Phospholipase D (2 units)	83
Phospholipase C (2 units) + 1% BSA	8
Phospholipase A (2 units) (detergent-solubilized membrane fraction)	39
None (delipidated membrane fraction)	24
Lipid (0.5 mg) (delipidated membrane fraction)	49
Lipid (1.0 mg) (delipidated membrane fraction)	46

Enzyme assays were performed as previously described[10,50].

lation activity in the presence of partially purified cytochrome *P*-450 and a crude lipid fraction (Table 3). Furthermore in affinity chromatographed preparations the 7.8 x 10⁻ -dalton polypeptide was enriched and in some instances it was the major band.

As evidenced by its visible spectrum, fluorimetric analyses and FMN stimulation, the flavin content of the affinity chromatographed reductase was substantially greater than of reductase obtained by ion exchange-gel filtration methods. In addition a significant increase in the recovery of reductase was observed when buffers used during the purification by affinity chromatography contained 1 μM FMN and FAD. The affinity-chromatographed reductase on fluorimetric analyses had 0.76 moles of FMN and 0.37 moles of FAD per mole of protein.

These results suggest a resemblance between plant and mammalian reductases which was further substantiated by studies of the physical and catalytic properties of the former. Upon thin layer isoelectric focusing of the reductase a pI of 5.3 was observed. The purified plant reductase transferred electrons to ferricyanide and dichlorophenolindophenol and exhibited menadione-mediated oxidase but not adrenaline oxidase activity. By kinetic analysis the apparent K_m for NADPH was determined to be 5.7 μM and that for cytochrome *c* to be 7.8 μM. The reductase was insensitive to antimycin A, dicoumarol, superoxide dismutase, and the alkaloid catharanthine (see below) while being inhibited by NADP (competitively, K_i = 18 μM), pCMB, and cetyltrimethylammonium bromide. One obvious difference between the plant *P*-450 reductase and its mammalian counterpart is the loose binding of the flavin prosthetic groups of the former. Of course, the facile loss of flavin may be attributable to the milieu the reductase finds itself in during homogenization and preliminary purification rather than a ready dissociation of holoenzyme.

Lipid Dependency. As shown in Table 4 phospholipases A and C inhibited the monoterpene hydroxylase to a significant extent whereas phospholipase D had no effect. This raises the possibility that the hydroxylase has a phospholipid requirement[55]. It is also plausible that free fatty acids or lysophospholipids released by the action of the phospholipases are inhibitory to the hydroxylase. To rule out this possibility the inhibition experiments were repeated in the presence of bovine serum albumin (BSA), which will adsorb these lipids. While BSA alone did not inhibit the hydroxylase (not shown), phospholipase C retained its inhibitory effect in the presence of this protein (Table 4). Suppression of hydroxylase activity by phospholipase A persisted in detergent-solubilized fractions, indicating the inhibition was not caused by disruption of membrane organization. Finally experiments were conducted in which after preincubation with phospholipase C the membrane fraction was sedimented by centrifugation to remove it from soluble phospholipase. Upon adding lipid (a $CHCl_3$:MeOH (2:1) extract of the 2 x 10^4 g pellet) back to the delipidated membrane about 50% of the activity was recovered (Table 4). These results suggest a phospholipid requirement for the hydroxylase. In conjunction with earlier studies, particularly on the reductase, the available data reveal a definite resemblance between the liver microsomal cytochrome P-450-dependent hydroxylase and the plant multienzyme complex.

Subcellular Localization. Initial studies on the hydroxylase were performed utilizing a microsomal 10^4 -10^5 g pellet[15]. But subsequent careful differential centrifugation under conditions designed to prevent aggregation of endoplasmic reticulum repeatedly revealed that approximately 60% of the total geraniol hydroxylase activity was associated with the 1 x 10^4 g pellet, whereas another 25-30% of the total activity was found in 1.5 - 2.0 x 10^4 g pellets[36]. To identify the membranes in the 10^5 g pellet and to define the role of the 58% of the total cytochrome P-450 associated with this fraction in *C. roseus*, its N-demethylase activity was investigated. Cytochrome P-450 dependent N-demethylases have been demonstrated to be localized in microsomes of castor bean endosperm on the basis of isopycnic density and marker enzyme studies[19]. N-Demethylase was found predominantly (70%) in the 10^5 g pellet of *C. roseus*, which showed that the latter contained the bulk of the endoplasmic reticulum.

These findings prompted an examination of the subcellular location of the hydroxylase by sucrose density gradient centrifugation, monitored by marker enzymes and electron microscopy. After a preliminary 3 x 10^3 g centrifugation, a 2 x 10^4 g pellet was obtained and separated on a discontinuous sucrose gradient. As seen in Figure 5 the hydroxylase was observed to occur in a single band with a density cen-

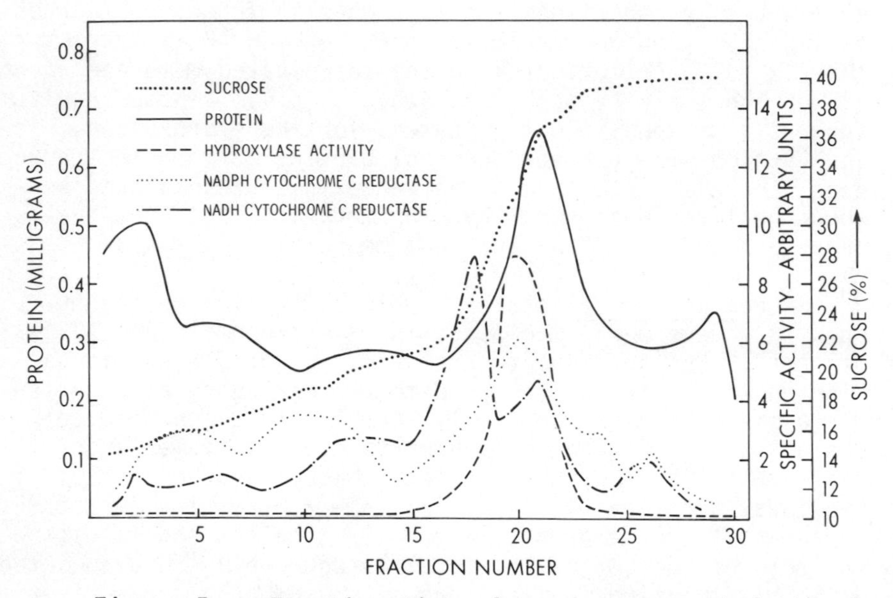

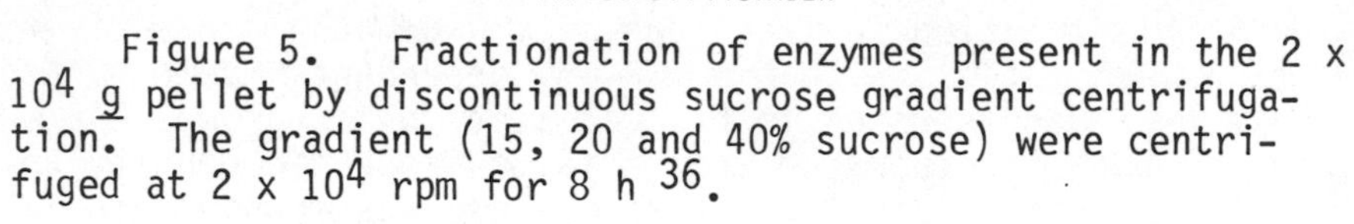
Figure 5. Fractionation of enzymes present in the 2 x 10^4 g pellet by discontinuous sucrose gradient centrifugation. The gradient (15, 20 and 40% sucrose) were centrifuged at 2 x 10^4 rpm for 8 h [36].

tered at 1.13 g/cm^3 which also gave the characteristic CO difference spectrum of reduced cytochrome *P*-450. Included in this yellow band were both NADH and NADPH cytochrome *c*-reductase and acid phosphatase activities. A mitochondrial enzyme marker (succinate dehydrogenase) was found exclusively at a density of 1.16 g/cm^3. The geraniol hydroxylase activity was restricted to a fraction with a density (1.13 g/cm^3) similar to that observed for a vacuolar fraction (provacuoles) obtained by Matile[56] from corn and tobacco seedlings (1.138 g/cm^3). Marker enzymes, esterase and acid phosphatase, were localized at the latter density in agreement with his findings. The bulk of the endoplasmic reticulum of castor bean endosperm is associated with a light membranous fraction having a density of 1.12 g/cm^3. This assignment was supported by isopycnic centrifugation in the presence and absence of Mg++. In the presence of Mg++ ribosomes are retained on the ER membrane, thereby affording a fraction with a density of 1.16 g/cm^3. The above experiments were repeated and although shifts in RNA and protein profiles were observed as reported[57], geraniol hydroxylase activity remained in the same fractions with the same density in both the presence and absence of 3 and 20 mM Mg++. Furthermore, the 10^5 *g* microsomal pellet was found to have an isopycnic density of 1.12 g/cm^3 in agreement with Beever's findings. Thus the geraniol hydroxylase was not associated with rough endoplasmic reticulum membranes.

In other approaches a 3 x 10^3 *g* supernatant of a cell-free homogenate was fractionated on a discontinuous gradient of 0.8, 1.0, 1.2 and 1.3 *M* sucrose in the presence of 20 m*M* EDTA and 30 m*M* Mg++ as described by Williamson *et al*[58]. In this manner four discrete bands were obtained of which the first, at the 0.8 *M* sucrose interface (1.10 g/cm^3), contained 90% of the total hydroxylase activity. Cotyledons of five-day-old etiolated seedlings were also macerated with pectinase[59] and vacuoles were obtained from the resultant protoplasts. Such vacuolar preparations contained specific activities six times that of the supernatant from which they were separated by centrifugation. Most definitive results were obtained when seedlings were ground in the presence of sand and in a medium containing 0.5 M sorbitol[60]. Upon subjecting the 2 x 10^4 *g* pellet preparation to a linear sucrose density gradient (15-50%) a threefold enrichment in hydroxylase activity was observed in a yellow band centered at 1.09-1.10 g/cm^3 (Figure 6). Part of this band extended into

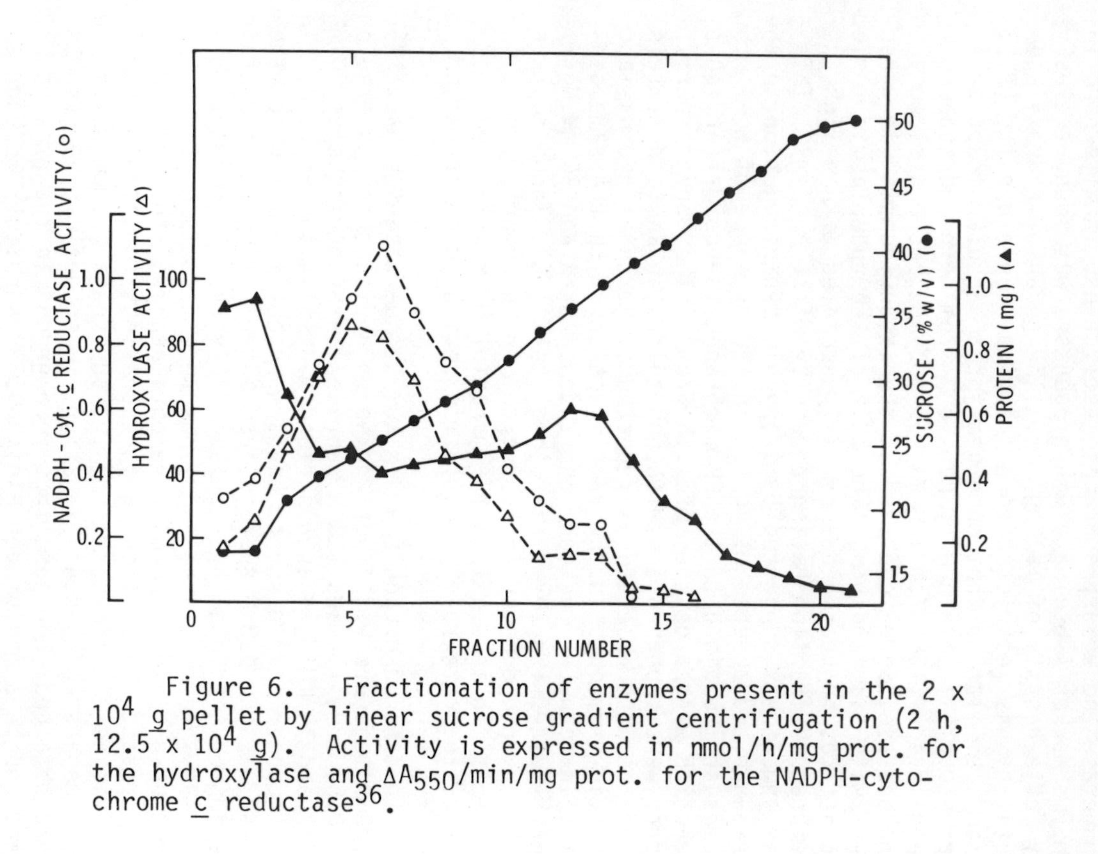

Figure 6. Fractionation of enzymes present in the 2 x 10^4 *g* pellet by linear sucrose gradient centrifugation (2 h, 12.5 x 10^4 *g*). Activity is expressed in nmol/h/mg prot. for the hydroxylase and ΔA_{550}/min/mg prot. for the NADPH-cytochrome *c* reductase[36].

the 1.13 g/cm^3 region and could correspond to the smaller hydroxylase fraction which was observed using 8h. sucrose density gradient centrifugation (Figures 5 and 6). Electron micrographs of the hydroxylase fraction of highest specific activity (1.09 g/cm^3) revealed a dense yellow vacuolar population ranging in size from 0.1-0.8 μm with the large vacuoles concentrated in the lower density fractions (Figures 7 and 8). Electron micrographs of the 1.13 g/cm^3 hydroxylase fractions from both preparations showed the presence of much smaller vesicles[36]. Vesicles containing a yellow pigment and similar in size to the large vacuoles were observed by light microscopy of free-hand sections of cotyledons. Examination by electron microscopy of fixed tissue of five-day-old etiolated cotyledons shows, in addition to a large central vacuole, coalescing provacuoles, which resemble the hydroxylase-containing vesicles in membrane thickness and size. These results suggest that the monoterpene hydroxylase, an enzyme participating in indole alkaloid biosynthesis is localized in vacuoles (provacuoles) of C. roseus.

There is a growing body of evidence that at least some plant alkaloids are restricted to specialized vesicles which are derived from endoplasmic reticulum[56,61,63]. Furthermore, these alkaloid vesicles appear to accumulate in laticifers. Perhaps the best support for this model has been gained in the study of morphine biosynthesis in Papaver somniferum. The plant contains characteristic capped vesicles which develop in laticifers, store alkaloids, and are found in preponderance in latex[61,64]. Some but not all of the enzymes of morphine biosynthesis may also be compartmentalized in these vesicles[65-67]. Since one of the extravesicular enzymes is a methyl transferase[67], it seems reasonable that some alkaloidal intermediates will be found outside the vesicles. Cytochemical data[62,63] support the ontogeny of vesicular membranes from endoplasmic reticulum by a dilation process.

Less is known about the cellular site of indole alkaloid biosynthesis in C. roseus. Recently, Yoder and Mahlberg[68] have obtained histochemical evidence for the localization of indole alkaloids in laticifers and specialized parenchyma cells in C. roseus by using alkaloid reagents. Such cells are far more abundant in mature tissue and were not detected in seedlings which were less than four days old. Older cotyledon and hypocotyl tissue of C. roseus contained these

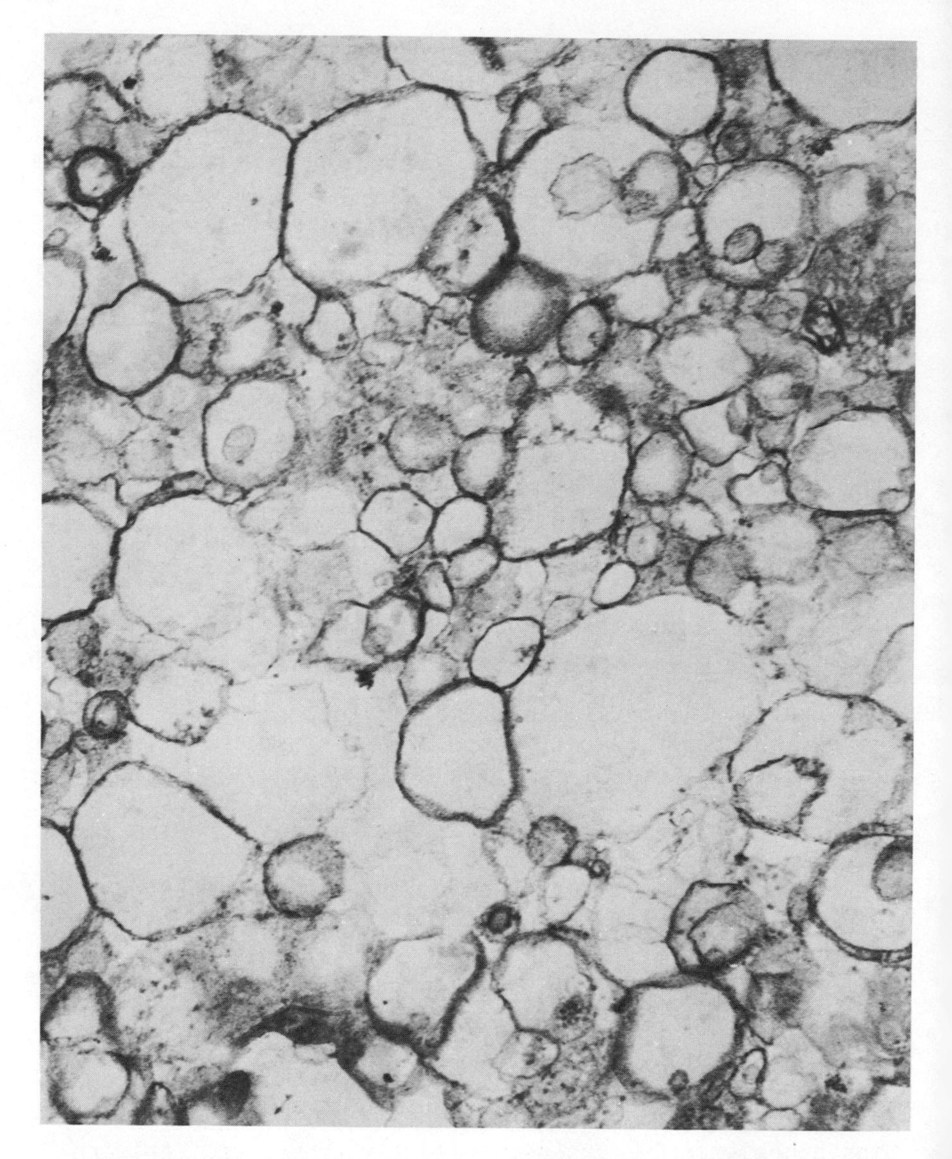

Figure 7. Electron micrograph of the pelleted yellow band obtained by sucrose gradient centrifugation following the procedure of Williamson _et al._[58]. The vesicles which contained 90% of the total hydroxylase activity ranged in size from 0.2-2 μm in diameter and 40-60 Å in membrane thickness. X 26,000 (reduced 40% for purposes of reproduction).

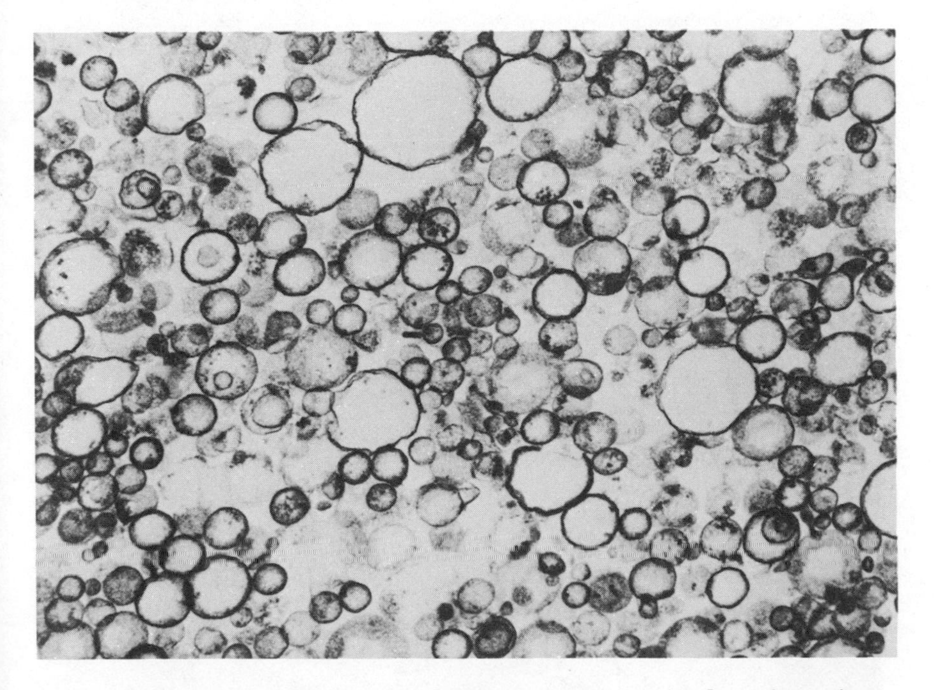

Figure 8. Electron micrograph of vesicular fractions having highest hydroxylase specific activities obtained by linear sucrose gradient centrifugation after homogenization in sorbitol[36]. X 28,000 (reduced 20% for purposes of reproduction).

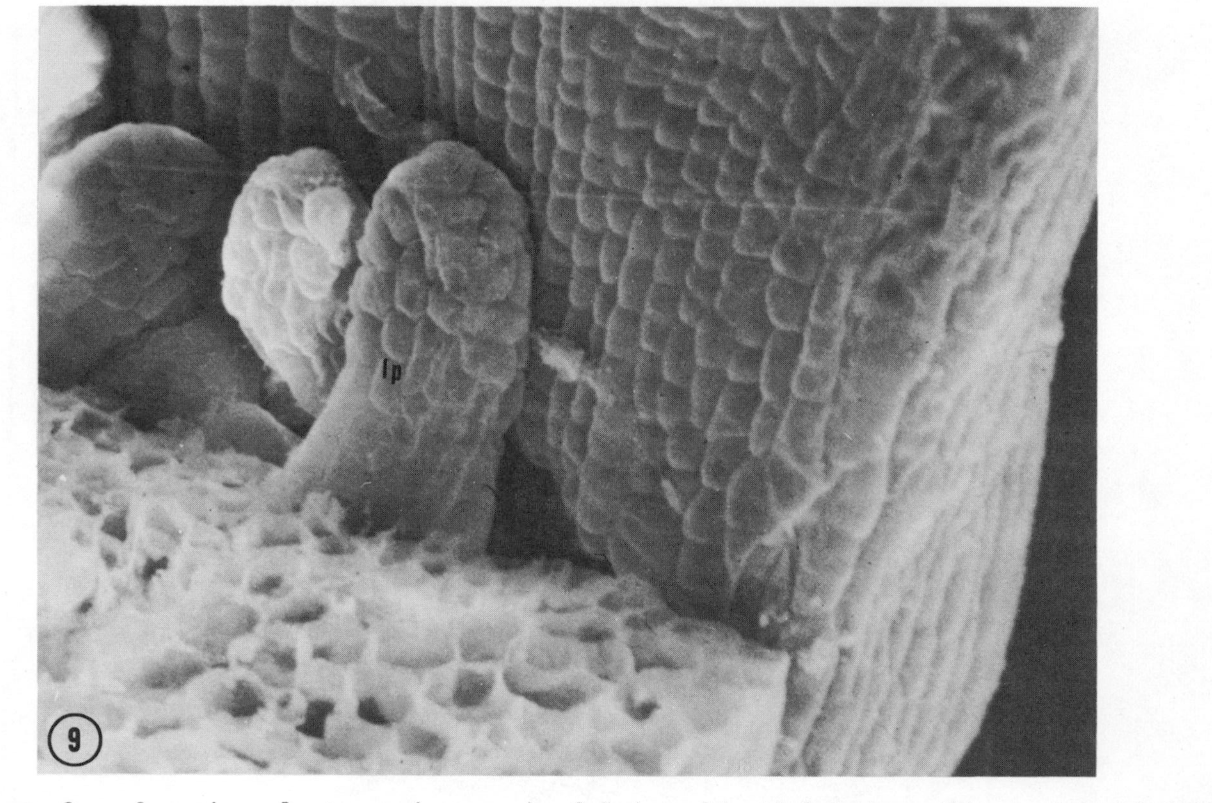

Figure 9. Scanning electron micrograph of 5-day-old etiolated seedling. Note that the apical meristem and leaf primordia (lp) comprise a minute fraction of the total tissue. X 65.

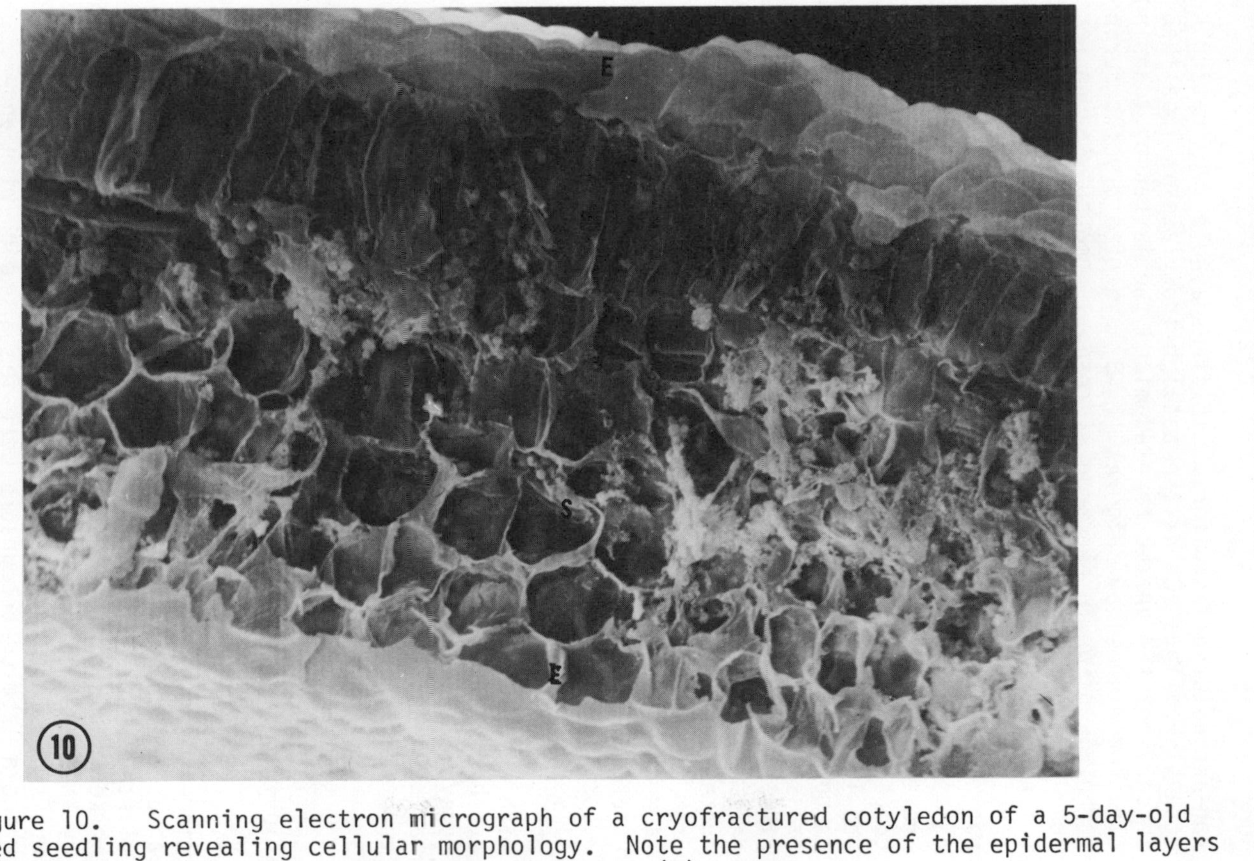

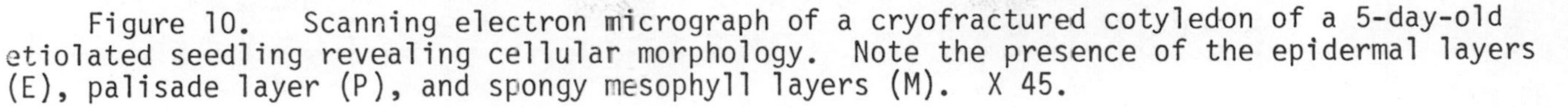

Figure 10. Scanning electron micrograph of a cryofractured cotyledon of a 5-day-old etiolated seedling revealing cellular morphology. Note the presence of the epidermal layers (E), palisade layer (P), and spongy mesophyll layers (M). X 45.

non-articulated laticifers. In five-day-old germinating C. roseus seedlings 90% of the total hydroxylase is localized in cotyledons[69] and we have examined the latter by scanning electron microscopy (Figure 9). By use of a cryofracturing technique prior to electron microscopy, the cotyledons were found to contain epidermal and parenchyma cells (Figure 10) but no laticifers at this age[70]. However, immature laticifers and their specialized parenchyma cell precursors may not be discernible without an alkaloid-sensitive cytochemical stain. It would be of interest to determine whether the provacuoles that contain the monoterpene hydroxylase are comparable to those observed in specialized parenchyma cells and laticifers. Unfortunately, C. roseus alkaloid vesicles do not appear to have a distinguishable characteristic such as is observed for those of P. somniferum. Attempts to use alkaloid detecting stains[71] for electron microscopy of intact C. roseus tissue have not been successful thus far and another approach was taken. Upon exposure of cotyledons from five-day-old etiolated seedlings to [^{3}H]vinblastine (4.8 μM, 10 μCi), which is synthesized in C. roseus, rapid uptake was observed. Examination of electron microscopic autoradiographs of the tissue revealed that 70% of the silver grains appeared over vacuoles (provacuoles) of mesophyll cells[70] (Figure 11). These results are consistent with the presence of a common site for at least part of the synthesis and storage of indole alkaloids in C. roseus. However, there is the reservation that exogenous alkaloids may be translocated to a subcellular locus that is different from the site where alkaloid molecules synthesized in situ.

Regulation. The above data on intracellular compartmentalization are compatible with regulation of indole alkaloid biosynthesis by feedback inhibition mechanisms. It has been well documented that amino acid biosynthesis in higher plants is regulated by end-product inhibition often reminiscent of that in bacterial systems[72]. Little is known about the control of alkaloid biosynthesis in higher plants[73]. Waller and his collaborators and Byerrum and coworkers have examined the interrelationship of ricinine biosynthesis and the pyridine nucleotide cycle in Ricinus communis[74]. In vivo tracer experiments have been carried out wherein labeled precursors were fed alone or with end-product alkaloids in an attempt to gain information on regulating effects of the latter. Similar experiments by Gross et al.[75] suggested that in Hordeum vulgare gramine was

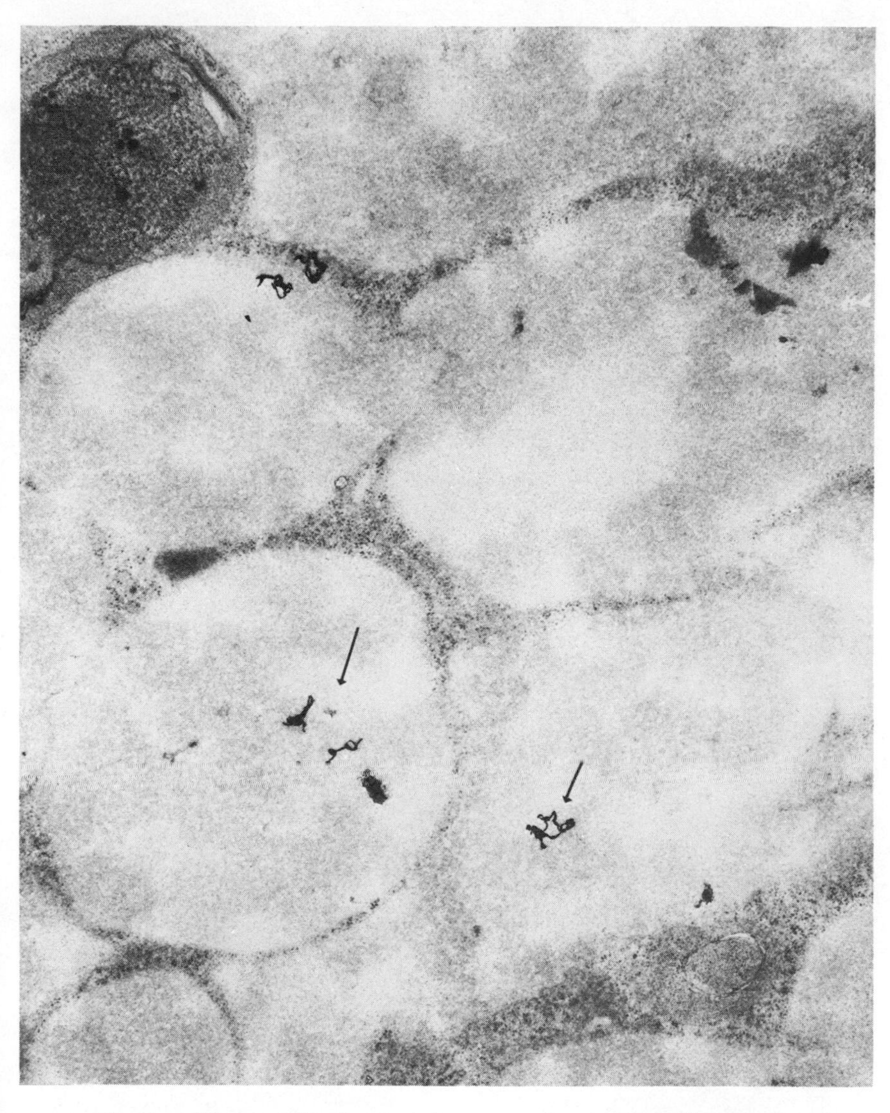

Figure 11. Electron microscopic autoradiograph of excised C. roseus cotyledon exposed to [^{3}H] vinblastine. Cotyledon slices were included with [^{3}H] vinblastine (4.8 μM, 10 μCi) for 3 h. Thereupon the cotyledons were washed exhaustively to remove exterior radioactivity and fixed for electron microscopic autoradiography[70]. X 25,000 (reduced 40% for purposes of reproduction).

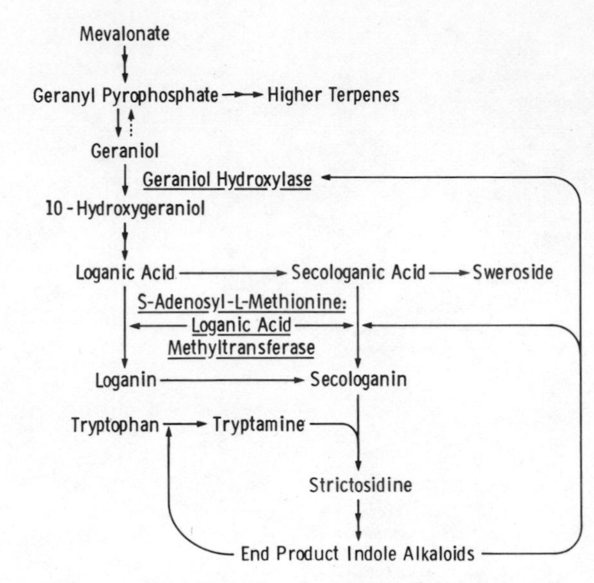

Figure 12. Hypothetical control points for feedback inhibition of indoele alkaloid biosynthesis.

Table 5. Effect of Multiple Alkaloid Addition on Geraniol Hydroxylase Activity

	% Activity Coadditive (0.5 mM)					
Alkaloid (0.5 mM)	None	Ajm.	Vind.	Vinb.	Cath.	Per.
Ajm.	87	86	82	108	62	77
Vind.	84	82	--	98	62	84
Vinb.	109	108	98	--	74	80
Cath.	62	62	62	74	--	57

Enzyme assays were carried out as previously described using a $2x10^4$ g pellet from *C. roseus*[10,50].

Abbreviations: Ajm. = Ajmalicine; Vind. = Vindoline; Vinb. = Vinblastine; Cath. = Catharanthine; Per. = Perivine.

either causing feedback inhibition or a stimulation of its own degradation. Obviously enzyme work is required as a sequel to the above experiments.

Baxter and Slaytor[76a] obtained evidence for feedback inhibition of tryptophan decarboxylase from *Phalaris tuberosa* by *N*,-*N*-dimethyltryptamine, an end-product alkaloid in the plant. Nicotine may repress three enzymes in *Nicotiana tabacum* involved in the synthesis from ornithine but no feedback inhibition was observed[76b]. Floss *et al*. have recently summarized the above findings on the regulation of alkaloid biogenesis as well as their own studies on ergot alkaloid control in the fungus *Claviceps purpurea*[73].

Since the monoterpene hydroxylase may catalyze the first irreversible step in indole alkaloid biosynthesis (see above) it may be one of the points of bifurcation betwen primary and secondary metabolism which is susceptible to feedback inhibition by end-product alkaloids (Figure 12). Preincubation of the monoterpene hydroxylase with representative alkaloids of the several families found in *C. roseus* showed that only catharanthine inhibited the enzyme[77]. Kinetic analyses revealed the alkaloid to be a reversible linear inhibitor (K_i = 1 nM) noncompetitive with geraniol and NADPH as substrates. Comparable inhibition of the solubilized hydroxylase by catharanthine tends to exclude a mechanism based upon disruption of membrane organization. Four other alkaloids, which alone were not inhibitory toward the hydroxylase, were also tested in the presence of a second alkaloid (Table 5). In branched anabolic pathways resembling that of indole alkaloid biosynthesis, there is precedent for cumulative and/or synergistic feedback inhibition. However, with the combinations used in Table 5 only catharanthine seemed to have a pronounced effect. Perivine alone at 0.25, 0.5 and 0.75 m*M* was without effect on the hydroxylase as were concentrations of vindoline as high as 2 m*M*[10]. The K_i of catharanthine inhibition (1 mM) is in the range of reported concentrations of this alkaloid in 3-4-month old *C. roseus* plants on a fresh-weight basis (0.3-1 mM)[78,79]. If compartmentalization of monoterpene hydroxylase and alkaloid occurs in a common vacuolar site, as suggested by the subcellular studies, then concentrations of catharanthine with respect to the hydroxylase may approach even higher values.

Table 6. Monoterpene Alcohol Oxidoreductase Activity in _C. roseus_

Substrate	Conditions*	Pyridine Nucleotide	Cpm x 10^{-3}(% Tot. Act.)‡ Aldehydes		Geraniol		Nerol	
$[1\text{-}^3H]$Geraniol	be	none	0.9	(3)	32.1	(94)	1.3	(4)
	c	none	1.0	(2)	47.7	(95)	1.6	(3)
	c	$NADP^+$	1.4	(9)	13.8	(88)	0.6	(4)
	c	NAD^+	6.0	(21)	27.1	(75)	1.3	(4)
	c	NAD^+, $NADP^+$	11.8	(26)	31.4	(69)	2.1	(5)
	c	NADH	1.1	(2)	46.5	(94)	1.8	(4)
	c	NADH, NADPH	0.9	(2)	33.4	(92)	1.9	(5)
$[1\text{-}^3H]$Geranial	be	none	55.7	(99)	0.3	(0.4)	0.6	(0.9)
	c	none	44.0	(99)	0.3	(0.6)	0.2	(0.6)
	c	NAD^+	50.1	(99)	0.4	(0.7)	0.3	(0.5)
	c	$NADP^+$	54.2	(99)	0.4	(0.6)	0.4	(0.6)
	c	NADH	27.1	(57)	18.8	(39)	2.0	(4)
	c	NADPH	34.4	(66)	15.7	(30)	1.8	(3)
$[1\text{-}^3H]$Neral	be		23.2	(95)	0.1	(0.5)	0.6	(2)
	c		6.8	(74)	0.7	(7)	1.4	(15)
$[1\text{-}^3H]$Nerol	be		2.7	(3)	2.7	(3)	78.1	(91)
	c		16.1	(15)	11.8	(1)	70.8	(66)

In 1.5 ml of 0.1 M potassium phosphate buffer, pH 7.2; 1 _mM_ DTT; 0.5 μCi $[1\text{-}^3H]$geraniol, nerol, geranial, or neral (30 μg), dissolved in 50 μl of acetone, 0.1 _mM_ NAD^+, $NADP^+$, NADH or NADPH; and _C. roseus_ extracts (100,000 x _g_ supernatant), 2 mg protein[69,80].

be = boiled enzyme; c = complete system.

A 10,000 x _g_ supernatant together with a combination of 0.1 M NAD^+, $NADP^+$, NADH and NADPH was used in these assays.

‡% Total Activity represents the percentage of total activity found on the TLC chromatoplate which was located in that spot. Since the workup of these very volatile compounds invariably caused losses due to volatilization, this internal representation of activity is probably more accurate.

MONOTERPENE ALCOHOL OXIDOREDUCTASE

There are enzyme activities of a more nonspecific type which may participate in indole alkaloid biosynthesis. The 10^5 *g* supernatant of *C. roseus* cell-free extracts catalyzes the oxidation of geraniol and nerol to geranial and neral respectively in the presence of oxidized pyridine nucleotides (NAD, NADP)[69,80]. Similarly the reverse reaction occurs when NADPH or NADH is added to incubation mixtures (Table 6). Alcohol dehydrogenase activity was assayed by a radiometric method after separation of geometric isomers by argentation TLC or by analytical GC.

Alcohol dehydrogenase activity was not observed in fractions sedimenting at 10^5 *g* or less. However, this does not preclude the occurrence of a soluble alcohol dehydrogenase within a vacuole (provacuole). Previous *in vivo* tracer studies have revealed that in the biosynthesis of cyclopentano and secocyclopentano monoterpene glucosides, the enzymes involved in the conversion of mevalonate to geranyl pyrophosphate exhibit a high stereospecificity[3,81]. Using asymmetrically labeled mevalonates as precursors, labeling patterns in monoterpene glucosides and alkaloids were obtained that were consistent with a stereochemical course of the earlier stages similar to that in mammalian isoprenoid biosynthesis[82]. Many subsequent steps of the indole alkaloid pathway exhibit stereospecificity. However, in the conversion of geranyl pyrophosphate to loganin at least two elements of stereospecificity are absent. First, *cis*-*trans* isomers are equally good substrates both *in vivo* and *in vitro*. Secondly an unusual randomization (for isoprenoid biosynthesis) of the terminal dimethyl groups of the acyclic monoterpene alcohols occurs. The latter event appears to take place only when both methyl groups are oxidized before glucosidation of the cyclopentano species[83]. Of the various hypotheses put forth to account for the *cis*-*trans* isomerization of acyclic monoterpenes we would invoke the previously mentioned redox mechanism. Including the initial work by Archer *et al.* in *Hevea*[82] and extensive studies in Arigoni's lab[84] the bulk of evidence[66,85-87] for a variety of plant systems seems to support the redox mechanism over two other possibilities cited in the literature.

Of the latter, one proposes that the prenyl transferases in *Pinus* and *Citrus* species synthesize both geranyl and neryl pyrophosphate despite the removal of the pro-R proton at C-2 of isopentenyl pyrophosphate in the formation of each[88]. This result may be explained by the hypothesis that unlike the mammalian prenyltransferase this plant enzyme does not extract different protons in forming cis and trans isoprenoids. In addition, Francis et al.[89] have shown that 4R tritiated mevalonates are converted to both tritiated geraniol and nerol in rose petals. Their interpretation of their findings is that geranyl pyrophosphate is formed and hydrolyzed to geraniol, which isomerizes to nerol by a redox mechanism. A similar explanation can be given for the data of Cori et al.[88] if the pyrophosphorylation step is reversible. There has been another report[90] that geranyl pyrophosphate and geraniol are converted to neryl pyrophos-phate and nerol, respectively, by a soluble enzyme system from carrot and peppermint in the presence of flavin, thiol or sulfide, and light. Although we have not detected direct isomerization of geraniol to nerol with either 2 x 10^4 g pellet, 10^5 g pellet, or supernatant from C. roseus and no interconversion of geranyl and neryl pyrophosphates was observed with cell-free extracts from orange flavedo[91], conditions of incubation were different than those used in the experiments with carrot and peppermint[90].

To explain the second element of stereospecificity that is missing, i.e., the randomization in the original terminal two methyl groups, we postulate the following sequence of events (Figure 13). An alcohol dehydrogenase may oxidize 10-hydroxygeraniol or 10-hydroxynerol to the corresponding aldehyde, which being conjugated like geranial or neral may be isomerized enzymically or non-enzymically.* This would give rise to a monoterpene alcohol with its 9-carbon at the aldehyde oxidation level. Since our specificity studies

* We prefer an enzymic mechanism since this would obviate non-enzymic labilization of the protons at C-5 of geraniol. The C-5 hydrogens are derived from C-2 of mevalonate and we have observed retention of the pro-R hydrogen at C-2 of mevalonate in loganic acid with the elimination of the pro-S[3]. Non-enzymic labilization of C-5 of geranial would not be consistent with our data. The same argument applies to the isomerization of the 10-oxogeraniol. Arigoni and Escher[5,66] have demonstrated that the identity of C-5 protons of mevalonate is also retained in this pathway, eliminating labilization of hydrogens at C-6 of geraniol.

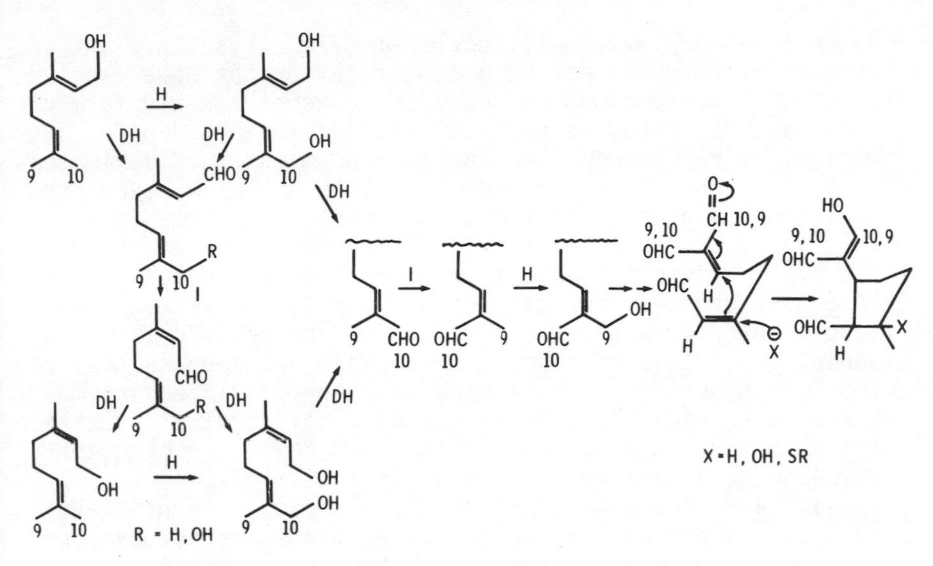

Figure 13. Hypothetical pathway for the conversion of geraniol and nerol to cyclopentano monoterpenes. H=hydroxylase, DH=dehydrogenase, and I=isomerase.

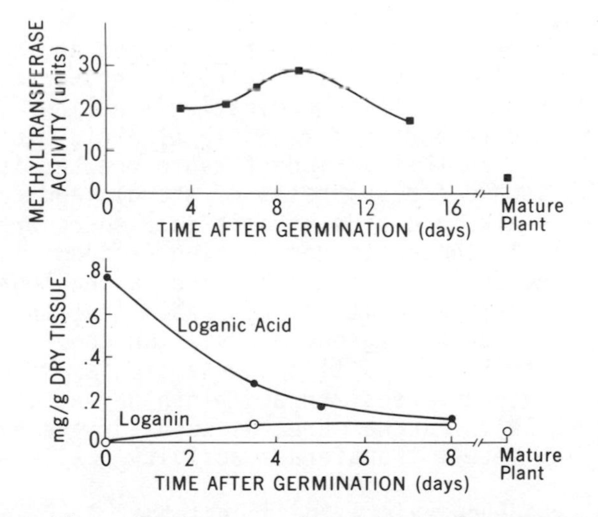

Figure 14. Variation of S-adenosyl-L-methionine: loganic acid activity with age of *C. roseus*[95]. Note parallel change in loganic acid and loganin content.

suggest hydroxylation will occur at C-10, this new 9-oxo-geraniol (or nerol) may be a substrate for the same cytochrome _P_-450-dependent hydroxylase. Once the newly formed hydroxymethyl group at C-10 is oxidized to an aldehyde, isomerization may reoccur leading to randomization.* Ultimately we propose that a trialdehyde undergoes a Michael type condensation to afford the first cyclopentano intermediate. This is supported by Leete's data in which labeled iridodial was not incorporated into loganin or indole alkaloids of _C. roseus_[92], and Inouye's data[83] in which 11-methyl cyclopentano monoterpenes such as deutzioside are derived from iridodial in _Deutzia crenata_. Furthermore, it appears that at least in some plants which make 11-methyl cyclopentano monoterpenes no randomization of the original terminal dimethyl groups occur. Thus, even if the 11-methyl is subsequently oxidized to a carboxyl, as in asperuloside in _Lamium amplexicaule_, no randomization was observed, the pathway in such plants proceeding via iridodial and 11-methyl cyclopentano monoterpene glucosides. On the other hand, in _C. roseus_ and related species, 11-methyl cyclopentanomonoterpeneglucoside derivatives have not been detected, while the occurrence of randomization is well established[1].

_S_-ADENOSYL-_L_-METHIONINE: LOGANIC ACID METHYLTRANSFERASE

Upon discovering that seeds of _C. roseus_ and _Strychnos nux vomica_ contain loganic acid in high concentrations (_ca._ 1%), it occurred to us that germinating seedlings could make good use of this precursor for indole alkaloid biosynthesis (see Figure 1). In this period of rapid growth, isoprenoid units are in demand for synthesis of sterols and carotenoids and yet indole alkaloid biosynthesis is also occurring relatively rapidly.* Indeed as germination proceeds loganic acid levels decline while a slight rise in the loganin pool size occurs[94]. Likewise it is reasonable that the enzyme responsible for carboxyl group methylation should be present in high concentrations and time course studies reveal maximal activity between the sixth and ninth day after germination (Figure 14). Furthermore, as seen for the hydroxylase, seedling methyl transferase activity was 7-8-fold as

* Excellent evidence exists to suggest that _C. roseus_ seeds are devoid of alkaloids but their synthesis commences within 29h. after germination[87,93].

great as that of mature plants. By acetone precipitation, ammonium sulfate fractionation, and DEAE-cellulose chromatography, a partially purified enzyme was obtained and characterized[95]. The apparent K_m for loganic acid was 12.5 m*M*, whereas for *S*-adenosyl-*L*-methionine it was 0.06 m*M*. Substrate specificity studies revealed loganic acid and secologanic acid were methylated at comparable rates but no activity was observed when 7-deoxyloganic acid or geniposidic acid[96] was used as substrate. This suggested that the enzyme has an essential binding site for a 7-hydroxyl group. Secologanic acid can satisfy this requirement because of its flexibility and the fact that the aldehyde moiety exists in a hydrated form. We thus prepared 7-epiloganic acid and found that it was not methylated. As can be seen with Dreiding models of loganic acid and its 7-epimer, although both can fill a putative active site where both 7-hydroxy and carboxy groups bind, the orientation of the glucosyl residue is completely different for each. Thus either the glucosyl residue also has an obligatory binding site or steric repulsion of the glucosyl group and amino acid residues occurs in the case of the 7-epimer. In fact 7-epiloganin is utilized by *Daphniphyllum macropodum* in asperuloside biosynthesis but only after epimerization to loganin with concomitant loss of the C-7 hydrogen[97].

[*O*-methyl-^{3}H]7-Deoxyloganin has been reported to be a precursor of loganin and indole alkaloids in *C. roseus* and degradation revealed specific labeling[98]. This data may be reconciled with the enzyme studies by postulating the existence of another methyltransferase which may be more important in mature plants. However, we have not been able to detect 7-deoxyloganin either by large-scale isolation or by isotope dilution analysis. A compound has been found that behaves chromatographically somewhat like 7-deoxyloganin but which with care can be resolved from it[80]. The facts that loganic acid is present in large quantities in *C. roseus* and that loganin has been shown to be readily incorporated into the alkaloids of *C. roseus*, all argue for the importance of the isolated methyltransferase. Thus enzymatic studies suggest that 7-hydoxylation precedes methylation in *C. roseus* at least during germination. Finally the methylation of secologanic acid affords an explanation of the incorporation of sweroside into indole alkaloids, again by way of a redox mechanism[94].

STRICTOSIDINE SYNTHETASE, β-GLUCOSIDASE, AND CATHENAMINE REDUCTASE

The Pictet-Spengler condensation is a common feature in the biosynthesis of a diversity of alkaloids in higher plants. Enzymes catalyzing such reactions are of considerable interest not only on the basis of their novelty but for mechanistic reasons as well. Recently two groups have isolated cell-free preparations capable of catalyzing the conversion of tryptamine and secologanin to ajmalicine and related *Corynanthe* alkaloids (Figure 15). Their elegant studies exemplify the advantage of the enzymatic approach to the study of metabolic pathways. Scott *et al.*[99] used a cell-free preparation from *C. roseus* seedlings and callus tissue whereas Zenk and coworkers[14] utilized *C. roseus* cell suspension cultures as their source of enzyme. By including a β-glucosidase inhibitor, D-δ-gluconolactone, in the system Stockigt and Zenk[100] were able to identify strictosidine as the first product of the Pictet-Spengler condensation. Earlier *in vivo* tracer studies[1] had implicated its 3-epimer, vincoside, as the key intermediate in indole alkaloid biosynthesis but the cell-free studies were contradictory. Subsequent *in vivo* and *in vitro* results have now clarified details of this part of the pathway and established strictosidine as the obligate intermediate for monoterpene indole alkaloids with both 3*R* and 3*S* configurations[101,102]. FAD was initially found to stimulate synthetase activity in a 3.8×10^4 *g* supernatant from seedlings and callus tissue[13]. On further purification[14,99], only reduced pyridine nucleotide was required for synthesis in ajmalicine and its epimers. In the absence of NADPH or NADH another intermediate, cathenamine (Figure 14) was formed[103]. Large-scale enzyme incubations afforded enough product to permit assignment of its structure. This crude enzyme system has been further characterized with an alkaloid-sensitive radioimmunoassay[104] which should prove to be a particularly powerful tool for plant biochemistry because of its sensitivity and rapidity. Thus the system appears to consist of the strictosidine synthetase, a β-glucosidase, and an NADPH-dependent (NADH is a less effective coenzyme) reductase. Gel filtration studies suggest *C. roseus* has an ajmalicine synthetase complex of molecular weight of 5.5×10^4, which is relatively low for a multienzyme system[18].

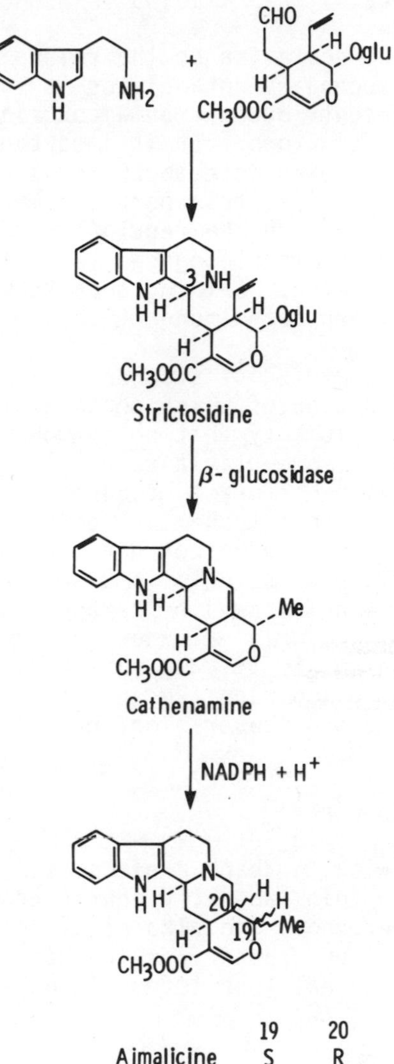

	19	20
Ajmalicine	S	R
19-Epiajmalicine	R	R
Tetrahydroalstonine	S	S

Figure 15. Enzymatic steps in the synthesis of ajmalicine and epimers from tryptamine and secologanin[14,93,99,100,104].

PHYSIOLOGICAL ROLE(S) OF ALKALOIDS IN HIGHER PLANTS

Considerable speculation on the raison d'etre of alkaloids and other secondary metabolites is invariably included in reviews. Our contribution would contain only equally speculative considerations. It is important to realize however that until we learn more about the subcellular and cellular sites of synthesis, transport and metabolism of secondary metabolites and the regulation of these processes, true insight into the physiological role of natural products will not be forthcoming. Gaining such knowledge requires the enzymological approach among others.

Because of the myriad of secondary metabolites in higher plants (only a fraction of which contain alkaloids) one is faced with the possibility that <u>no</u> common function may prevail. For example, some play a role which may be designated as hormonal but with different mechanisms and different results. Speculation on allochemical physiological roles of secondary metabolites may be correct but only in other cases. In the cycle of scientific discovery correct interpretations are often followed by recognition of exceptions and refinement of the original knowledge. Perhaps an anticipation of such a possibility in the study of secondary metabolism will not discourage the investigator, but will prevent dogmatism and promote original experimentation.

ACKNOWLEDGEMENTS

The authors wish to thank Professors D. Arigone, J. M. Bobbitt, P. Heinstein, and R. C. Hutchinson for generous samples of monoterpenoid and alkaloidal metabolites. This work was supported by grants from the National Science Foundation and the National Institutes of Health.

REFERENCES

1. Cordell, G. A. 1974. The biosynthesis of indole alkaloids. Lloydia 37:219-298.
2. Leete, E. 1977. Alkaloid biosynthesis. Specialist Periodical Reports, Biosynthesis, 5:136-239.
3. Coscia, C. J., L. Botta and R. Guarnaccia. 1970. Monoterpene biosynthesis III. On the mechanism of iridoid and secoiridoid monoterpene biosynthesis. Arch. Biochem. Biophys. 136:498-506.
4. Battersby, A. R., S. H. Brown and T. G. Payne. 1970. Biosynthesis of loganin and the indole alkaloids from hydroxygeraniol-hydroxynerol. J. Chem. Soc., Chem. Commun. 827-828.
5. Escher, S., P. Loew and D. Arigoni. 1970. The role of hydroxygeraniol and hydroxynerol in the biosynthesis of loganin and indole alkaloids. J. Chem. Soc., Chem. Commun. 823-826.
6. Sumner, J. B. 1926. The isolation and crystallization of the enzyme urease. J. Biol. Chem. 69:435-441.
7. Leete, E. 1969. Alkaloid biosynthesis. Adv. Enzymol. 32:373-422.
8. Loomis, W. D. 1974. Overcoming problems of phenolics and quinones in the isolation of plant enzymes and organelles. Meth. Enzymol. 31A:528-544.
9. Madyastha, K. M., T. D. Meehan and C. J. Coscia. 1976. Characterization of cytochrome P-450 dependent monoterpene hydroxylase from the higher plant V. rosea. Biochemistry 15:1097-1102.
10. McFarlane, J., unpublished observations.
11. Patterson, B. D. and D. P. Carew. 1969. Growth and alkaloid formation in Catharanthus roseus tissue cultures. Lloydia 32:131-140.
12. Baxter, C., unpublished results.
13. Scott, A. I., and S. L. Lee. 1975. Biosynthesis of the indole alkaloids. A cell-free system from Catharanthus roseus. J. Am. Chem. Soc. 97:6906-6908.
14. Stockigt, J., J. Treimer, M. H. Zenk. 1976. Synthesis of ajmalicine and related indole alkaloids by cell-free extracts of Catharanthus roseus cell suspension cultures. FEBS Lett. 70:267-270.
15. Meehan, T. D. and C. J. Coscia. 1973. Hydroxylation of geraniol and nerol by a monooxygenase from Vinca rosea. Biochem. Biophys. Res. Commun. 53:1043-1048.

16. Hasson, E. P. and C. A. West. 1976. Properties of the system for the mixed function oxidation of kaurene seed of Marah macrocarpus. Plant Physiol. 58:479-484.
17. Murphy, P. J. and C. A. West. 1969. The role of mixed function oxidases in kaurene metabolism in Echinocystis macrocarpa Greene endosperm. Arch. Biochem. Biophys. 133:395-407.
18. Frear, D. S., H. R. Swanson and F. S. Tanaka. 1969. N-Demethyalation of substituted 3-(phenyl)-1-methylureas: isolation and characterization of a microsomal mixed function oxidase from cotton. Phytochemistry 8:2157-69.
19. Young, O. and H. Beevers. 1976. Mixed function oxidases from germinating castor bean endosperm. Phytochemistry 15:379-385.
20. Benveniste, I. and M. F. Durst. 1974. Mise en evidence dans les tissus de tubercule de topinambour d'une enzyme a cytochrome P-450, l'acide trans-cinnamique 4-hydroxylase. C. R. Acad. Sci. Paris 278:1487-1490.
21. Benveniste, I., J. Salaun, and F. Durst. 1977. Wounding-induced cinnamic acid hydroxylase in Jerusalem artichoke tuber. Phytochemistry 16:69-73.
22. Potts, J. R. M., R. Weklych and E. E. Conn. 1974. The 4-hydroxylation of cinnamic acid by sorghum microsomes and the requirement for cytochrome P-450. J. Biol. Chem. 249:5019-5026.
23. Russell, D. W. 1971. The metabolism of aromatic compounds in higher plants. J. Biol. Chem. 246:3870-3878.
24. Croteau, R. and P. E. Kolattukudy. 1975. Biosynthesis of hydroxy fatty acid polymers. Enzymatic epoxidation of 18-hydroxyoleic acid to 18-hydroxy-cis-9,10-epoxystearic acid by a particulate preparation from spinach. Arch. Biochem. Biophys. 170:61-72.
25. Soliday, C. L. and Kolattukudy, P. E. 1978. Midchain hydroxylation of 16-hydroxypalmitic acid by the endoplasmic reticulum fraction from germinating Vicia faba. Arch. Biochem. Biophys. 188:338-347.
26. Markham, A., G. C. Hartman and D. V. Parke. 1972. Spectral evidence for the presence of cytochrome P-450 in microsomal fractions obtained from some higher plants. Biochem. J. 130:90P.
27. Rich, P. R. and D. S. Bendall. 1975. Cytochrome components of plant microsomes. Eur. J. Biochem. 55:333-341.

28. Rich, P. R. and Lamb, C. J. 1977. Biophysical and enzymological studies upon the interaction of trans-cinnamic acid with higher plant microsomal cytochromes P-450. Eur. J. Biochem. 72,353-360.
29. Dus, K., personal communication.
30. Cohn, J. A., A. P. Alvares and A. Kappas. 1977. On the occurrence of cytochrome P-450 and aryl hydrocarbon hydroxylase activity in rat brain. J. Exptl. Med. 145:1607-1611.
31. Galloway, M. P., unpublished observations.
32. Ishimaru, A. and I. Yamazaki. 1977. The carbon monoxide-binding heme protein reducible by hydrogen peroxide in microsomal fractions of pea seeds. J. Biol. Chem. 252:199-204.
33. Cotte-Martinon, M. G., V. Yahiel and G. Ducet. 1974. Induction d'un cytochrome du type P-450 et de peroxydase durant la survie du tubercule de pomme de terre. Phytochemistry 13:2085-2090.
34. Yahiel, V., M. G. Cotte-Martinon and G. Ducet. 1974. Un cytochrome de type P450 dans le spadice d'Arum. Phytochemistry 13:1649-1651.
35. Lembi, C. A. and D. J. Morre. 1970. Isolation of plasma membrane-rich cell fractions from onion stem and corn coleoptiles. Plant Physiol. 46(Suppl.):14.
36. Madyastha, K. M., J. E. Ridgway, J. G. Dwyer and C. J. Coscia. 1977. Subcellular localization of a cytochrome P-450-dependent mono-oxygenase in vesicles of the higher plant Catharanthus roseus. J. Cell. Biol. 72:302-313.
37. Morrow, C., unpublished results.
38. Licht, H. J. and C. J. Coscia. 1978. Cytochrome P-450LM$_2$ mediated hydroxylation of monoterpene alcohols Biochemistry, in press.
39. Coscia, C. J. and R. Guarnaccia. 1968. Natural occurrence and biosynthesis of a cyclopentanoid monoterpene carboxylic acid. J. Chem. Soc., Chem. Commun. 138-140.
40. Licht, H. J., unpublished observations.
41. Perez, L. M., L. Chavet, M. de la Fuente, M. C. Rojas, G. Portilla, U. Hashagen, L. A. Fernandez and O. Cori. 1978. Biosynthesis of mono- and sesquiterpenoids by soluble enzymes from Citrus flavedo. Abst. Papers Joint Meet. Am. Soc. Pharmacog. Phytochem. Soc. of North Am., 1st, Stillwater, Oklahoma, August 14-17, 1978:18.

42. Christophe, J. and G. Popjak, 1961. Studies on the biosynthesis of cholesterol: XIV. The origin of prenoic acids from allyl pyrophosphates in liver enzyme systems. J. Lipid Res. 2:244-257.
43. Tsai, S. C. and J. L. Gaylor. 1966. Testicular sterols. V. Preparation and partial purification of a microsomal prenol pyrophosphate pyrophosphohydrolase. J. Biol. Chem. 241:4043-4050.
44. Baisted, D. J. 1967. Incorporation of label from geraniol-^{14}C into squalene, β-amyrin and β-sitosterol in germinating pea seeds. Phytochemistry 6:93-97.
45. Botta, L. 1968. Ph.D. Dissertation No. 4098 Eidg. Tech. Hochschule Zurich, Switzerland. Zur Biogenese von Verbindungen der Lupanreihe.
46. van Aller, R. T. and W. R. Nes. 1968. The phosphorylation of geraniol in germinating peas. Phytochemistry 7:85-88.
47. Madyastha, K. M. and W. D. Loomis. 1969. Phosphorylation of geraniol by cell-free enzymes from Mentha piperita. Fed. Proc. 28:665.
48. Seubert, W. and E. Fass. 1964. Untersuchungen uber den bakteriellen Abbau von Isoprenoiden. Biochem. Z. 341:23-34.
49. Seubert, W. and U. Remberger. 1963.. Untersuchungen uber den bakteriellen Abbau von Isoprenoiden. Biochem. Z. 338:245-264.
50. Madyastha, K. M. and C. J. Coscia. 1974. Detergent-solubilized NADPH-cytochrome c (P-450) reductase from the higher plant, Catharanthus roseus: Purification and characterization. J. Biol. Chem. (in press).
51. Vermilion, J. L. and Coon, M. J. 1978. Purified liver microsomal NADPH-cytochrome P-450 reductase. J. Biol. Chem. 253:2695-2704.
52. Aoyama, Y., Y. Yoshida, S. Kubota, H. Kumaoka and A. Furumichi. 1978. NADPH-cytochrome P-450 reductase of yeast microsomes. Arch. Biochem. Biophys. 185:362-369.
53. Fan, L. L. and B. S. S. Masters. 1974. Properties of purified kidney microsomal NADPH-cytochrome c reductase. Arch. Biochem. Biophys. 165:665-671.
54. Yasukochi, Y. and B. S. S. Masters. 1976. Some properties of a detergent-solubilized NADPH-cytochrome c (cytochrome P-450) reductase purified by biospecific affinity chromatography. J. Biol. Chem. 251:5337-5344.

55. Buche, T. and H. Sandermann. 1973. Lipid dependence of plant microsomal cinnamic acid 4-hydroxylase. Arch. Biochem. Biophys. 158:445-447.
56. Matile, P. 1966. Enzymes of vacuoles from rootlets of corn seedlings. A contribution to the functional significance of vacuoles to intracellular digestion. Z. Naturforsch. 21b:871-878.
57. Lord, J. M., T. Kagawa, T. S. Moore and H. Beevers. 1973. Endoplasmic reticulum as the site of lecithin formation in castor bean endosperm. J. Cell Biol. 57:659-667.
58. Williamson, F. A., D. J. Morre and M.J. Jaffe. 1975. Association of phytochrome with rough-surfaced endoplasmic reticulum fractions from soybean hypocotyls. Plant Physiol. 56:738-743.
59. Takebe, I., Y. Otsuki and S. Aoki. 1968. Isolation of tobacco mesophyll cells in intact and active state. Plant Cell Physiol. 9:115-124.
60. Matile, Ph. and A. Wiemken. 1974. Vacuoles and spherosomes. Meth. Enzymol. 31A:572-578.
61. Fairbairn, F. W., F. Hakim, and Y. E. Kheir. 1974. Alkaloidal storage, metabolism and translocation in the vesicles of Papaver somniferum latex. Phytochemistry 13:1133-1139.
62. Nessler, C. L. and P. G. Mahlberg. 1976. Laticifers in stamens of Papaver somniferum L. Planta 129:83-85.
63. Nessler, C. L. and P. G. Mahlberg. 1977. Ontogeny and cytochemistry of alkaloidal vesicles in laticifers of Papaver somniferum L. (Papaveraceae). Amer. J. Bot. 64:541-551.
64. Dickenson, P. B. and J. W. Fairbairn. 1975. The ultrastructure of the alkaloidal vesicles of Papaver somniferum latex. Ann. Bot. 39:707-712.
65. Antoun, M. D. and M. F. Roberts. 1975. Enzymic studies with Papaver somniferum. 5. The occurrence of methyltransferase enzymes in poppy latex. Planta Medica 28:6-11.
66. Escher, S. 1972. Stereochemische Aspekte der Biosynthase von Indolalkaloiden. Thesis Dissertation No. 4887. Eidg. Tech. Hochschule, Zurich, Switzerland.
67. Roberts, M. F. and M. D. Antoun. 1978. The relationship between L-DOPA decarboxylase in the latex of Papaver somniferum and alkaloid formation. Phytochemistry 17:1083-1087.

68. Yoder, L. R. and P. G. Mahlberg. 1976. Reactions of alkaloid and histochemical indicators in laticifers and specialized parenchyma cells of Catharanthus roseus (apocynaceae). Am. J. Bot. 63:1167-1173.
69. Meehan, T. D., unpublished observations.
70. Ridgway, J. E., unpublished results.
71. Neumann, D. and E. Muller. 1967. Intracellular detection of alkaloids in plant cells by light and electron microscopic criteria. Flora 158:479-491.
72. Bryan, J.K. 1976. Amino acid biosynthesis and its regulation, in plant Biochemistry. 3rd ed. Eds. J. Bonner and J. E. Varner, Academic Press, New York, 525-560.
73. Floss, H. G., J. E. Robbers and P. F. Heinstein. 1976. Regulatory control mechanism in alkaloid biosynthesis. Rec. Adv. Phytochem. 8:141-178.
74. Waller, G. R. and E. K. Nowacki. 1978. Alkaloid biology and metabolism in plants. Plenum Press, New York, New York.
75. Gross, D., H. Lehman and H. P. Schutte. 1970. Zur Physiologie der Graminbildung. Z. Pflanzenphysiol. 63:1-9.
76a. Baxter, C. and M. Slaytor. 1972. Partial purification and some properties of tryptophan decarboxylase from Phalaris tuberosa. Phytochemistry 11:2763-2766; Biosynthesis and turnover of N,N-dimethyltryptamine and 5-methoxy-N,N-dimethyltryptamine in Phalaris tuberosa. Ibid, 2767-2773.
76b. Mizusaki, S., Y. Tanabe, M. Noguchi, and E. Tamaki. 1973. Phytochemical studies on tobacco alkaloids XVI. Changes in the activities of ornithine decarboxylase, putrescine N-methyltransferase and N-methyl-putrescine oxidase in tobacco roots in relation to nicotine biosynthesis. Plant Cell Physiol. 14:103-110.
77. McFarlane, J., K. M. Madyastha and C. J. Coscia. 1975. Regulation of secondary metabolism in higher plants. Effect of alkaloids on a cytochrome P-450 dependent monoxygenase. Biochem. Biophys. Res. Commun. 66:1263-1269.
78. Daddona, P. E., J. L. Wright and C. R. Hutchinson. 1976. Alkaloid catabolism and mobilization in Catharanthus roseus. Phytochemistry 15:941-945, and personal communication.
79. Leete, E., A. Ahmad and I. Kompis. 1965. Biosynthesis of the Vinca alkaloids. I. Feeding experiments with

tryptophan-2-C^{14} and acetate-1-C^{14}. J. Amer. Chem. Soc. 87:4168-4174.
80. Guarnaccia, R., unpublished observations.
81. Battersby, A. R., J. C. Byrne, R. S. Kapil, J. A. Martin, T. G. Payne, D. Arigoni and P. Loew. 1968. The mechanism of indole alkaloid biosynthesis. J. Chem. Soc., Chem. Commun. 951-953.
82. Archer, B. L., D. Barnard, E. G. Cockbain, J. W. Cornforth, R. H. Cornforth and G. Popjak. 1966. The stereochemistry of rubber biosynthesis. Proc. Roy. Soc. B163:519-523.
83. Inouye, H., S. Ueda and S. Uesato. 1977. Zum Mechanismus der Methylcyclopentan-gerustbildung bei der Biosynthese einiger Iridoidglucoside. Tetrahedron Lett. 709-712; and Uber die Biosynthese des Deutziosids, ibid 713-716.
84. Arigoni, D. 1975. Stereochemical aspects of sesquiterpene biosynthesis. Pure Appl. Chem. 41:219-245.
85. Banthorpe, D. V., B. M. Modawi, I. Poots and M. G. Rowan. 1978. Redox interconversions of geraniol and nerol in higher plants. Phytochemistry 17:1115-1118.
86. Chayet, L., R. Pont-Lezica, C. George-Nascimento and O. Cori. 1973. Biosynthesis of sesquiterpene alcohols and aldehydes by cell free extracts from orange flavedo. Phytochemistry 12:95-101.
87. Mothes, K., I. Richter, K. Stolle and D. Groger. 1965. Physiologische Bedingungen der Alkaloid-synthese bei Catharanthus roseus G. Don. Naturwissenschaften 52:431.
88. Jedlicki, E., G. Jacob, F. Faini, O. Cori and C. A. Bunton. 1972. Stereospecificity of isopentenylpyrophosphate isomerase and prenyl transferase from Pinus and Citrus. Arch. Biochem. Biophys. 152:590-596.
89. Francis, M. J. O., D. V. Banthorpe and G. N. J. Le Patourel. 1970. Biosynthesis of monoterpenes in rose flowers. Nature 228:1005-1006.
90. Shine, W. E. and Loomis, W. D. 1974. Isomerization of geraniol and geranyl phosphate by enzymes from carrot and peppermint. Phytochemistry 13:2095-2101.
91. George-Nascimento, C. and O. Cori. 1971. Terpene biosynthesis from geranyl and neryl pyrophosphates by enzymes from orange flavedo. Phytochemistry 10:1803-1810.

92. Bowman, R. M. and E. Leete. 1969. Observations on the administration of iridodial-7-^{14}C to Vinca Rosea. Phytochemistry 8:1003-1007.
93. Scott, A. I., P. B. Reichardt, M. B. Slaytor and J. G. Sweeny. 1971. Mechanisms of indole alkaloid biosynthesis. Recognition of intermediacy and sequence by short-term incubation. Bioorg. Chem. 1:157-173.
94. Guarnaccia, R., L. Botta and C. J. Coscia. 1973. Biosynthesis of acidic iridoid monoterpene glucosides in Vinca rosea. J. Am. Chem. Soc. 96:7079-7084.
95. Madyastha, K. M., R. Guarnaccia, C. Baxter and C. J. Coscia. 1973. Monoterpene biosynthesis VII: S-Adenosyl-L-methionine: Loganic acid methyltransferase, a carboxyl-alkylating enzyme from Vinca rosea. J. Biol. Chem. 248:2497-2501.
96. Guarnaccia, R., K. M. Madyastha, E. Tegtmeyer and C. J. Coscia. 1972. Monoterpene biosynthesis VI: geniposidic acid, an iridoid glucoside from Genipa americana. Tetrahedron Lett. 5125-5127.
97. Inouye, H., S. Ueda and Y. Takeda. 1969. Loganin als Prekursor in der Biosynthese des Asperulosids. Z. Naturforsch. 24b:1666-1667.
98. Battersby, A. R., A. R. Burnett and P. G. Parsons. 1970. Preparation and isolation of deoxyloganin: its role as precursor of loganin and the indole alkaloids. J. Chem. Soc., Chem. Commun. 826.
99. Scott, A. I., S. L. Lee and W. Wan. 1977. Indole alkaloid biosynthesis: partial purification of "ajmalicine synthetase" from Catharanthus roseus. Biochem. Biophys. Res. Commun. 75:1004-1009.
100. Stockigt, J. and M. H. Zenk. 1977. Isovincoside (strictosidine), the key intermediate in the enzymatic formation of indole alkaloids. FEBS Lett. 79:233-237.
101. Brown, R. T., J. Leonard and S. K. Sleigh. 1978. The role of strictosidine in monoterpenoid indole alkaloid biosynthesis. Phytochemistry 17:899-900.
102. Rueffer, M., N. Nagakura and M. H. Zenk. 1978. Strictosidine, the common precursor for monoterpenoid indole alkaloids with 3α and 3β configuration. Tetrahedron Lett. 1593-1596.
103. Stockigt, J., H. P. Husson, C. Kan-Fan and M. H. Zenk. 1977. Cathenamine, a central intermediate in the

cell free biosynthesis of ajmalicine and related indole alkaloids. J. Chem. Soc., Chem. Commun. 164-166.

104. Treimer, J. F. and M. H. Zenk. 1978. Enzymic synthesis of corynanthe-type alkaloids in cell cultures of *Catharanthus roseus*: quantitation by radioimmunoassay. Phytochemistry 17:227-231.

Chapter Five

FROM TERPENES TO STEROLS: MACROEVOLUTION AND MICROEVOLUTION

GUY OURISSON, MICHEL ROHMER, AND
ROBERT ANTON

Institut de Chimie
Universite Louis Pasteur, Strasbourg and
Laboratoire de Matiere Medicale
Faculte de Pharmacie
Universite Louis Pasteur, Strasbourg

INTRODUCTION

We shall address ourselves in this chapter to one central question of comparative biochemistry, with obvious phytochemical overtones: is it possible to discern the precise biochemical mechanisms by which biochemical evolution can take place? We will propose a rather detailed description of such pathways at two levels: on the one hand, we will discuss present evidence related to the macroevolutionary events leading to sterols; that is, related to the development from prebiotic systems of complex substances universally present in eukaryotic cells; on the other, at the lower level of microevolution, we will apply some of the hypotheses developed at the higher level to offer a detailed scheme for the infrageneric variability of sterol-like triterpenes in one special tissue: the latex of Euphorbiaceae.

Our discussion will be mainly based on experimental results obtained in Strasbourg in four apparently unrelated fields: organic geochemistry, microbial biochemistry, cell culture biochemistry, and plant chemotaxonomy. We have of course also used evidence given by other workers; however, whereas it is usually easy to quote accurately the experimental results called upon to illustrate our discussion, it is often much more difficult to give proper credit to those who have helped to give shape to the theoretical aspects. Not that they are minor contributors to the development of our own ideas, but the fact is that their influence may often be indirect, protracted, and unrecognized, and that the final outcome may bear little resemblance to the individual remarks which have nurtured it. We hope that our contribution does not only represent the sum of past bits of knowledge, but that it is qualitatively novel, that it is a breakthrough.*

* The foregoing had been written in July, 1978 and the manuscript submitted when we became aware of the review paper of Nes (Nes, W. R., 1974. Lipids, 9: 596) introducing the notion of "sterol-like" molecules, and of biochemical evolution of these substances as membrane components. The "equivalence" of sterols with tetrahymanol and "carotenols" is proposed. One of us (G.O.) had attended a lecture given by Dr. Nes on the general theme of sterols, sterol-like structures, membranes and evolution, but had not consciously remembered it. Being aware of cryptomnesia, the selective fading of memories still accessible subconsciously, he feels compelled to acknowledge that his interest in the present topic may have been triggered several years ago by Dr. Nes. However, even a superficial analysis of Nes' paper shows that our own views differ fundamentally from his so that plagiarism need not be taken into account.

We will first introduce some basic tenets of biochemical evolutionary theory, as we understand them. Evolutionary steps result from discrete mutations which happen randomly and are usually lost because of their incompatibility with the survival of the cell, the organism, or the population. A single mutation, at the biochemical level, can be equated with one change in nucleic acid sequences, that is, (leaving aside any redundancy of the genetic code) with one change in amino acid sequence in one protein. This molecular "evolutionary drift" occasionally leads to changes compatible with survival. When these changes bring about an advantage, compared with previous conditions, the mutants may become in a Darwinian position of progressive dominance; "advantage" in this context can be either intrinsic (biochemical, physiological) or extrinsic to the mutant organism (eclogical, *i.e.* operating only in the interactions with the surroundings). It may also be negatively advantageous: that is, it may result in the disappearance of a previous tolerable disadvantage. In contrast to Darwinian evolution, non-Darwinian mutations, neutral or even marginally disadvantageous, normally occur and are maintained. The flow of evolution is turbulent, not smooth, and eddies are compatible with its general progression. Furthermore, increasing variation is entropically favored.

Before embarking on our main discourse we will summarize a few facts to give a background on which we will try to build.

OCCURRENCE OF STEROIDS AND RELATED COMPOUNDS

Eukaryotes. Steroids are present in eukaryotes under an extraordinary variety of guises. Thus they occur as the vitamins or hormones of mammals or insects, the constituents of saponins and toxic glycosides of plants and marine organisms, and as the defense substances of water beetles, etc. There can be no common denominator to all these important functions. However, we shall consider them all to be *derived* from *one* key function, common to all eukaryotes: sterols (*sensu stricto*, *i.e.* the 3 β-alcohols of about 27-30 carbons) are present in every eukaryotic cell at the level of about 1 mg/g of dry weight and they are integral parts of the cell membranes. They are associated with the *n*-acylphospholipids (*e.g.* the lecithins) and act by virtue of their size, amphiphilic character and rigidity, as modifiers of the interactions between the straight chains of the lipid in the fluid state, to stabilize the double-layer structure of the hydrophobic part of the membrane[1].

Some eukaryotes are capable of endogenous sterol biosynthesis: plants, fungi and, among animals, at least the vertebrates. For the other eukaryotes which are unable to biosynthesize sterols the compounds are in fact vitamins; this is well documented for insects, but may be true of many other groups of animals.

Prokaryotes. In contrast to their wide distribution in eukaryotes, sterols are absent in prokaryotes, bacteria or cyanobacteria. While the bulk of the published evidence is overwhelmingly in favor of this statement[2-4], the presence of sterols in some prokaryotes is repeatedly reaffirmed. We have ourselves[5] recently shown that indeed, with modern analytical methods, sterols can be found in any prokaryote studied. However, the sterol mixture thus isolated consistently has the same composition (a mixture of cholesterol and phytosterols, such as can be obtained from the washings of the fingers of the experimenter and of laboratory rubber teats). Furthermore, the minute amount isolated is reduced as one increases precautions against contamination and addition of radioactive precursors such as [^{14}C]-acetate to selected prokaryotes has never led to their incorporation into the sterols, which must therefore be accidental contaminants.

The only prokaryotes containing sterols are the wall-free bacteria, the mycoplasmas, which acquire them from their environment (with the exception of *Acholeplasma laidlawii*, which can dispense with any cholesterol)[6]. However, one bacterium, *Methylococcus capsulatus*, contains, not sterols proper, but 4α-methylsterols[7]. We shall come back to this important case later.

Many prokaryotes, bacteria or cyanobacteria (about 50% of the strains studied), thus devoid of sterols, contain substances rather comparable to them in size, rigidity and amphiphilic character: the hopanoids[8-11] (Figure 1, Table 1). These triterpenes most probably play, in the membrane of organisms containing them, the same role as sterols in the eukaryotic membrane. At least in one case, that of an

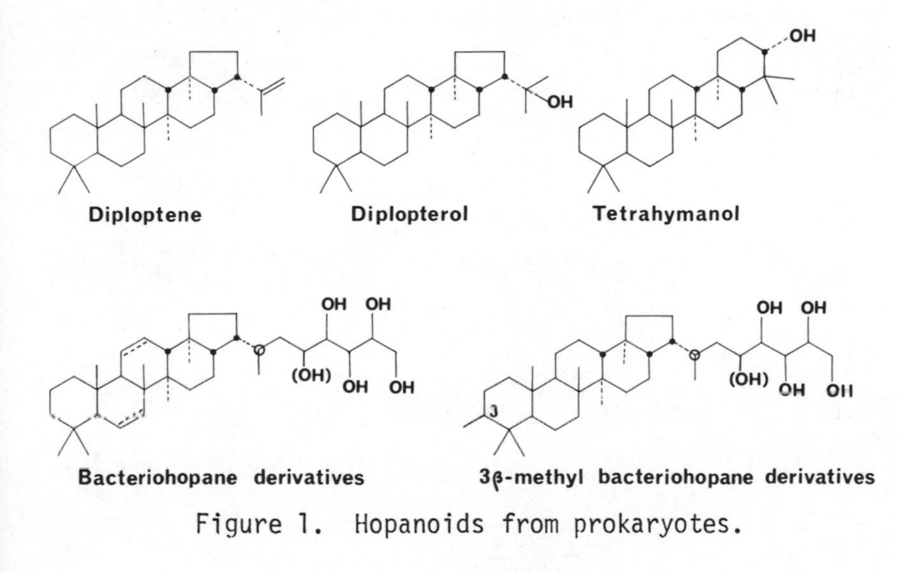

Figure 1. Hopanoids from prokaryotes.

Table 1. Distribution of Hopanoids in Prokaryotes

	Hopanopids not detected	Hopanoids present
Cyanobacteria	. *Synechococcus* sp. *Spirulina* sp.	. *Anabaena* sp. *Nostoc* sp. (2 strains) *Synechocystis* sp. (2 strains)
Purple sulfur bacteria	*Chromatium* sp. *Amoebobacter* sp. *Thiocapsa* sp.	
Green sulfur bacteria	*Chlorobium* (2 strains)	
Purple nonsulfur bacteria		*Rhodopseudomonas* (6 strains) *Rhodospirillum* *Rhodomicrobium*
Methylothrophs		*Methylococcus* (7 strains) *Methylomonas* *Methylocystis* *Methylosinus*
Other bacteria	*Thiobacillus* (2 strains)	
		Nitrosomonas europaea
	Pseudomonas fluorescens *Ps. aeruginosa* *Ps. stutzeri* *Ps. maltophila* *Ps. diminuta* *Xanthomonas campestris* *Rhizobium lupini* *Agrobacterium tumefaciens* *Caulobacter crescentus*	*Pseudomonas cepacia*
		Azotobacter vinelandii *Acetobacter* (12 strains, covering 9 species)
	Moraxella (2 strains) *Escherichia coli* *Proteus vulgaris*	
	Bacillus subtilis	*Bacillus acidocaldarius*
	Sporosarcinia lutea *Clostridium paraputrificum* *Streptococcus faecalis* *Micrococcus luteus*	
	Micromonospora sp.	*Streptomyces chartreusi*
	Actinoplanes brasiliensis *Desulfovibrio desulfuricans* *Methanobacterium thermoautotrophicum* *Halobacterium cutirubrum* *Sulfolobus acidocaldarius* *Thermoplasma acidophilum*	

eukaryote, however, the protozoon Tetrahymena pyriformis, the hopanoids (sensu lato, i.e. including as their major constitutent the gammacerane derivative tetrahymanol) are biosynthesized, and membrane-bound, in sterol-free cultures; they can be promptly replaced by sterols when these are supplied in the diet. Hopanoid biosynthesis is then repressed[12].

This leaves about 50% of the prokaryote strains studied in Strasbourg to be devoid both of sterols and of hopanoids[10]. We propose that the biophysical role of these rigid monopolar 15-20 A inserts can be played in other micro-organisms by other additives to the double-layer membrane (Figure 2): rigid 30-40 A monopolar inserts (monohydroxylated carotenoids), rigid bipolar inserts, or bracers of the same length (α,ω-dihydroxylated carotenoids), or nonrigid bipolar bracers (α,ω-diglyceryl ethers of a head-to-head dimer of phytanol (Figure 3). These reinforcement mechanisms are partly supported by experimental evidence, but most of this is so far indirect.

AN EVOLUTIONARY SCHEME FOR STEROLS

We have analyzed the possible biochemical phylogeny of sterols in the following way. We start with the assumption that the function sterols play in the eukaryotic membrane is an essential one in that their lack would be lethal to the organism, but we assume simultaneously that the precise molecular mechanism by which proper membrane reinforcement is achieved can be variable. We shall therefore proceed from sterols backwards, trying to discern a phylogeny of the other membrane strengtheners mentioned above.

From Hopanoids to Sterols. The immediate phylogenetic precursors of the sterols may have been their cellular equivalents, the hopanoids. Sterols and hopanoids are both products of the same general biosynthetic pathway, that to the terpenoids, such as the carotenoids and the other dimers of phytanol mentioned earlier. To accept hopanoids as having led to sterols by biochemical evolution, one would have to become convinced that (a) they are more primitive than sterols; and (b) their biosynthesis may have mutated to that of sterols. We believe that this is indeed the case. Hopanoids and sterols derive from the same acyclic

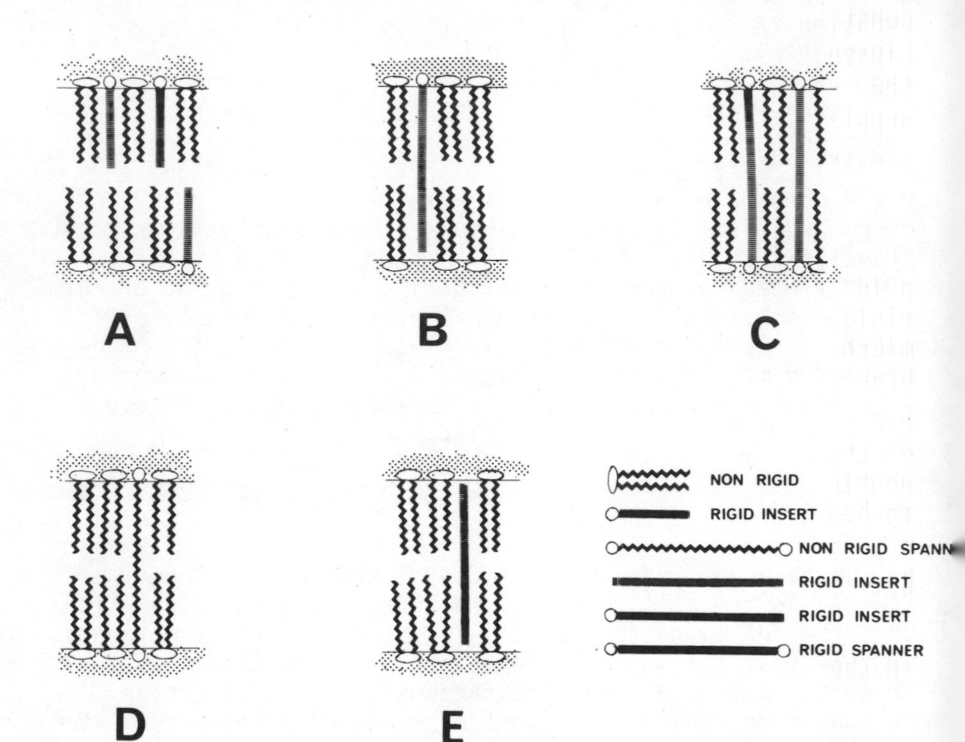

Figure 2. Hypothetic mechanisms of membrane stabilization. (A) Rigid amphiphilic inserts comparable to sterols or hopanoids. (B) Rigid amphiphilic inserts such as monoglucosides of carotenoids. (C) Rigid dipolar "spanners" such as diglucosides of carotenoids. (D) Nonrigid dipolar "spanners" such as bis(phytanyl) ethers. (E) Rigid hydrophobic inserts such as carotenes.

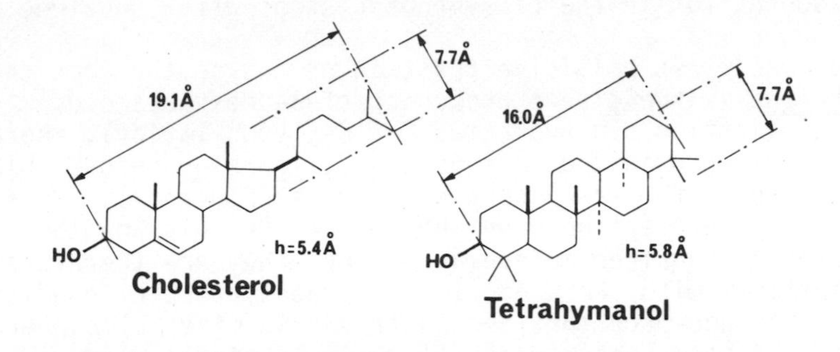

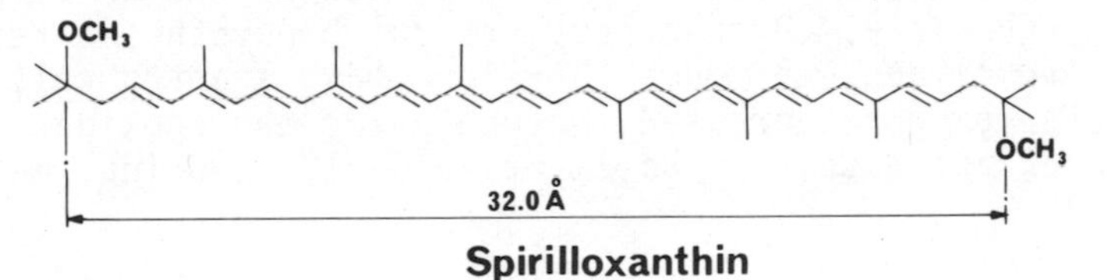

Figure 3. Comparative sizes of various potential membrane inserts.

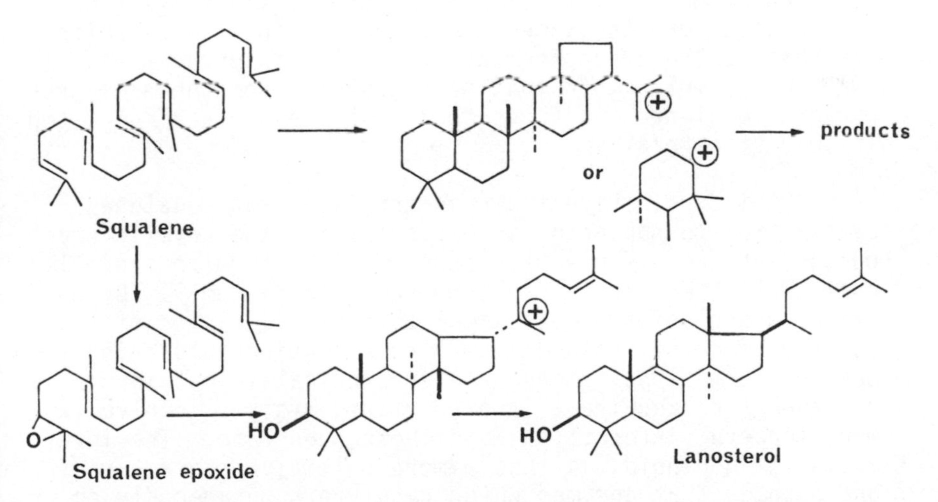

Figure 4. Comparison of biosynthesis of hopanoids and sterols.

C_{30}squalene. The "primitiveness" of hopanoids may therefore be sought for in the biosynthetic steps after squalene.

One first primitive trait is of course the fact that squalene is the direct precursor of 3-deoxyhopanoids[13-14], by cyclization and hydration (or C-glycosylation), whereas it is first oxidized to squalene 2,3-epoxide before this is cyclized to the precursors of sterols (Figure 4). In fact, none of the steps between the C_2 acetate unit and the hopanoids requires molecular oxygen; they are therefore all compatible with archaebiotic, prephotosynthetic, conditions, if it is accepted that the Earth's primimtive atmosphere was anaerobic and that almost all the molecular oxygen came from photosynthesis[15]. By contrast, sterol biosynthesis requires molecular oxygen, and must therefore have started only after molecular oxygen increased, not only for the epoxidation of squalene, but also for the later removal of nuclear methyl groups.

A second primitive trait is the fact that hopanoids are derived from squalene by a simple cyclization, without rearrangement[16]. By contrast, sterol biosynthesis requires cyclization of squalene epoxide to a protolanostane system, and subsequent 1,2-rearrangements of methyl groups and hydrogen atoms, to give either lanosterol (in fungi and vertebrates) or its isomer cycloartenol (in plants) which are then further transformed. Hopanoid biosynthesis is therefore simpler than that of sterols: from the same precursor, squalene, the function is obtained in a single step instead of several.

A third primitive trait is the fact that squalene cyclization to hopanoids involves folding the acyclic precursor into an all-prechair conformation, to give the all-chair, all-trans-anti-trans, pentacyclic system[16]. By contrast, sterol biosynthesis implies the cyclization of the acyclic precursor folded into a conformation comprising preboat conformations. These are energetically less favorable and therefore require of the enzymatic systems involved a more forceful interaction with their substrate. The formation of hopanoids is thus a more primitive character if one assumes that enzymes which catalyze spontaneously occur ring processes are more primitive than others which bring about reactions which are excluded (though in principle permissible) *in vitro*.

More direct information on the primitive character of the squalene-hopanoid cyclase, compared with the squalene-oxido-cyclase comes from our study of three cyclizing systems obtained from micro-organisms by cellular disintegration and fractionation.

In a cell-free system obtained from the bacterium *Acetobacter rancens*, which contains hopanoids, but no sterols, we have found that squalene is indeed incorporated into the pentacyclic triterpenes diplopterol and diploptene[17] (Figure 5). But, contrary to our expectations, squalene epoxide which is absent from the bacterium, is an acceptable substrate for the cyclase, and this, without selectivity for any of the enantiomers of the epoxide. One obtains therefore *in vitro* a mixture of the 3α- and 3β-hydroxy derivatives of diplopterol and diploptene. By contrast, the squalene-oxido-cyclase of eukaryotes displays a much stricter specificity, in that it acts only upon squalene 2,3 *S*-epoxide, and not on the 2,3 *R*-antipode nor on squalene itself[18]. We realize of course that for some enzymes biochemical evolution may well lead to progressively lesser selectivity; however, we feel safe in assuming that here lesser substrate selectivity is a primitive trait.

In the protozoon *Tetrahymena pyriformis*, an eukaryote which produces tetrahymanol and a little diplopterol when grown in the absence of sterols[12], a similar cell-free system displays a comparable lack of substrate specificity: squalene is cyclized to the two natural triterpenoids mentioned, but its (*RS*) 2,3-epoxide is also cyclized, and gives a mixture of 3α- and 3β-hydroxy derivatives of tetrahymanol and diplopterol[5].

The bacterium *Methylococcus capsulatus* was known to produce hopanoids (diploptene[7], but mostly C_{35} bacteriohopane derivatives[11]), and 4 -methylsterols. From this microorganism, we have similarly obtained by cellular disruption and fractionation a cell-free system with cyclase activity (Figure 6). From squalene, it produces exclusively hopanoids. However, when fed (*R*,*S*) squalene epoxide, it gives a mixture of the same 3α- and 3β-hydroxydiploptenes,

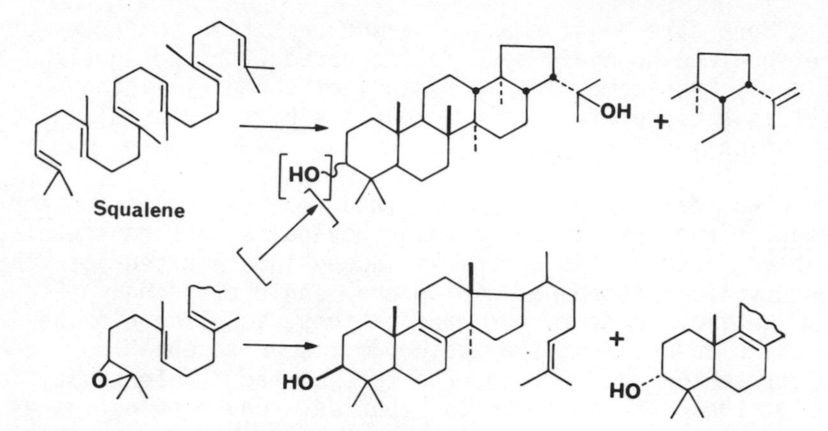

Figure 5. Biosynthesis of hopanoids in a cell-free system of Acetobacter rancens.

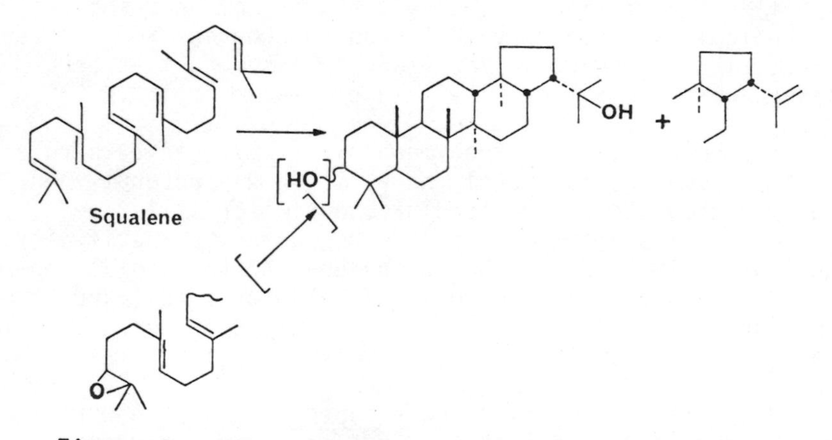

Figure 6. Biosynthesis of hopanoids and lanosterol in a cell-free system from Methylococcus capsulatus.

and of lanosterol and 3-epilanosterol[5]; in this case, like in all others, oxidative steps (2,3-epoxidation of squalene, angular methyl oxidative removal) cannot be observed in our *in vitro* preparations. We consider this result as highly informative. It implies that, in *Methylococcus*, *two* cyclases are present. The first one, producing hopanoids, is identical to, or resembles, the *Acetobacter* cyclase. The second one, the lanostane one, is substrate-specific, in that it does not act on squalene, but only on its epoxide; however, it is not ("not yet" !) as highly substrate-specific as the cyclase of yeast or of mammals, since it acts on both enantiomers of the epoxide. We have therefore, in this micro-organism, simultaneously one "primitive" cyclase, and one more elaborate, but not ("not yet") as elaborate as its eukaryote analog. Furthermore, the absence of hopan-3β,22 -diol in the bacterium implies that the hopane-producing cyclase cannot act on the squalene epoxide, even though this must be present in the cell: there must be rigorous *compartmentation*. One does not yet know whether, as appears probable, 4α-methylsterols are used as membrane inserts in one part of the cell, and hopanoids in another.

To accept hopanoids as potential *phylogenetic* precursors of sterols, it is still necessary to make two further comments: first, to eliminate *other* similar molecules (say, other triterpenes) as equally probable precursors, and second, to show that a biochemical mechanism compatible with evolutionary tenets can be discerned to account for the change from hopanoids to sterols.

Triterpenes other than hopanoids are excluded as phylogenetic precursors of sterols for several reasons. First, many of them have "bent" structures, incompatible with an efficient fit with the n-acyl chains of the membrane phospholipids. Second, most of them derive from squalene epoxide (note therefore: they must have arisen in the post-photosynthetic era) by a mechanism as complex as for sterols, involving cyclization but also rearrangements (1,2--shifts of hydrogens, and methyl groups, and ring enlargement) or later oxidative changes neither of which can be regarded as a primitive trait[16]. Furthermore many of them are not amphiphilic, but are dihydroxy derivatives (*e.g.* the dammarenediols). The fact is, therefore, that hopanoids fulfill uniquely criteria characterizing potential sterol precursors.

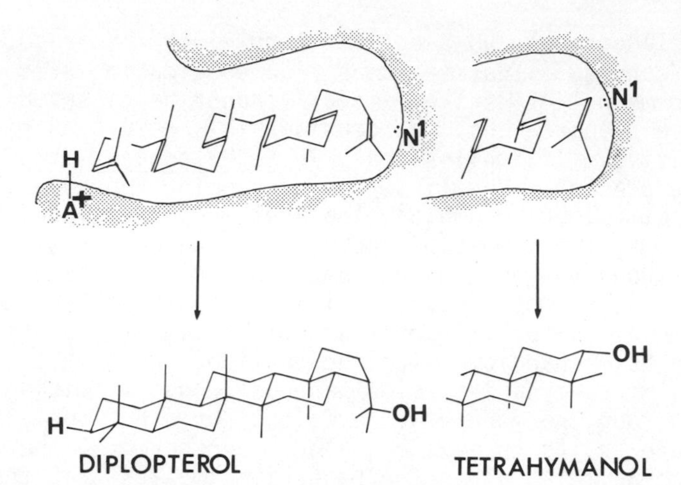

Figure 7. Schematic representation of the enzymatic biosynthesis of diplopterol and tetrahymenol. A^+-H represents an acidic center. $:N^1$ is a nucleophilic center (or a bound water molecule).

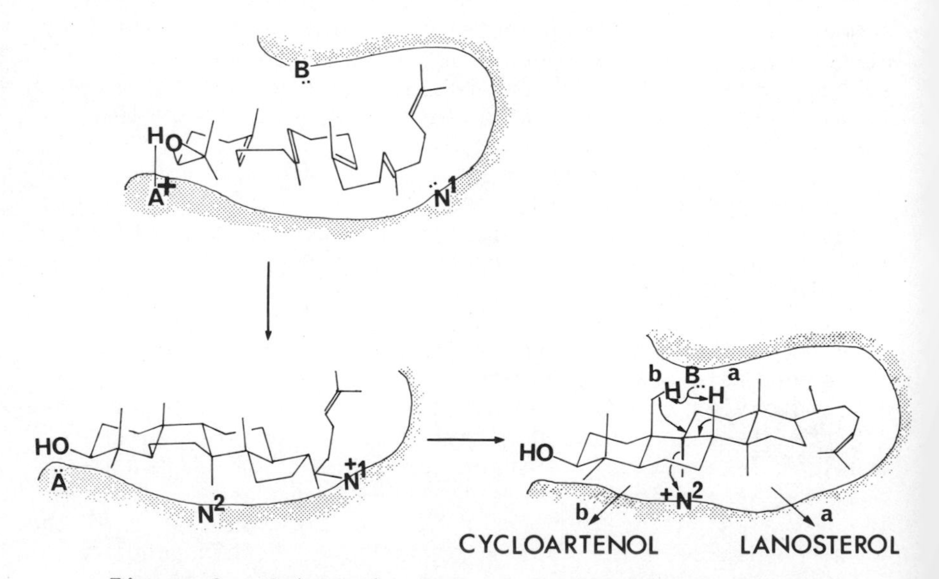

Figure 8. Schematic representation of the enzymatic biosynthesis of cycloartenol and lanosterol.

Mechanism of Change. Now, let us examine the mechanism of change; we have to admit that our complete ignorance of the structure of the cyclases does not help in understanding their evolution ! However, we can conjecture[19,20]. We shall accept that, whether monomolecular or complex, the enzymes (or enzymatic systems) responsible for the cyclization share the characteristic of carrying, in a precise relationship, an acidic center ($A-H^+$), and a nucleophilic or basic center N^1:. In the most primitive squalene cyclases (e.g. in Acetobacter) the steric contraints around $A-H^+$ would be small, so that squalene or its two diastereomeric 2,3-epoxides can be accommodated (Figure 7). A small change of position (less than 2 A) of the terminus N^1:, which can be a water molecule, could account for the first evolutionary step we postulate: from diplopterol to tetrahymanol. This change can easily be brought about by a one amino acid change in a protein sequence around N^1:. We assume that, in the phylum leading to Tetrahymena, two enzymes are present, similar yet different: the original one leading to diplopterol, and a mutant form, perhaps an isoenzyme, producing tetrahymanol.

To proceed further, towards sterols, one would have to assume that the nucleophilic $N:^1$ center has undergone also a small change of position, to enable it to bind to the terminus, C-20, of the initial tetracyclized product (Figure 8). Such a change would again entail at most a 3 A displacement easily compatible with a one amino acid mutation. However, this would have to be accompanied by a change in the shape of the cavity, forcing squalene epoxide into the pre chair-boat-chair conformation necessary for the formation of the protolanostane system. Such a small variation could well be a consequence of the first change and this is also possible for the functional specificity of the cyclase of Methylococcus acting on the squalene epoxide to give lanosterol. One can then link up these postulates with the hypothesis put forward earlier, to explain the formation of lanosterol (or cycloartenol, in plants[21]: rearrangement to a new intermediate bound by C-9 to a second nucleophilic center N^2:, and elimination of H-8 by a basic center B:. N^2: and B: may have been present in the original enzyme leading to hopanoids, and simply not operate there: however, see below). Of course, these consequences of a (postulated) single amino acid mutation would lead only to lanosterol, not to sterols. The later stages may well have been initially little specific, and have entailed general-function oxidases. We have no information on that part of the biosynthetic pathway from an evolutionary point of view.

In summary, hopanoids can be considered as having led to sterols by plausible structural changes of a single enzymatic system. We have discussed elsewhere the problems associated with the larger cross-section of hopanoids, as regards their hydrophobic interactions with the membrane phospholipids[10].

From Tetraterpenes to Hopanoids. We have already mentioned that hopanoids are present in only about one-half of the strains of prokaryotes studied by us (Table 1). In the other strains, we have assumed that the function played by sterols and hopanoids is still exerted by some constituents, and we have assumed that these can, at least in some cases, be identified with some carotenoids or other C_{40} tetraterpene derivatives[10]. We shall now show that these can indeed not only fulfill the function of membrane strengtheners, but also be phylogenetic precursors of hopanoids, and therefore of sterols.

The structures involved here are quite varied. They include essentially traditional C_{40} carotenoids, most of which are terminally hydroxylated and/or glucosylated. C_{50} carotenoids are also structurally compatible with their postulated function as spanners: they often carry in fact hydrophilic head-groups sufficiently far apart to reach through a normal membrane. C_{30} carotenoids, derived from squalene, are also known in bacteria; they would be too short to span the membrane, but their presence may be the sign that the "normal" primitive metabolism leading to C_{40} carotenoids requires little change to produce the precursor demanded for triterpene biosynthesis.

It would not be realistic to speak of "primitive traits" in the biosynthesis of carotenoids, however, for it implies reactions simpler, in a way, than the stereospecific pentacyclization of squalene to hopanoids. Not enough is known of the distribution of specific bacterial carotenoids, nor of their association with phospholipids, to warrant any further comment.

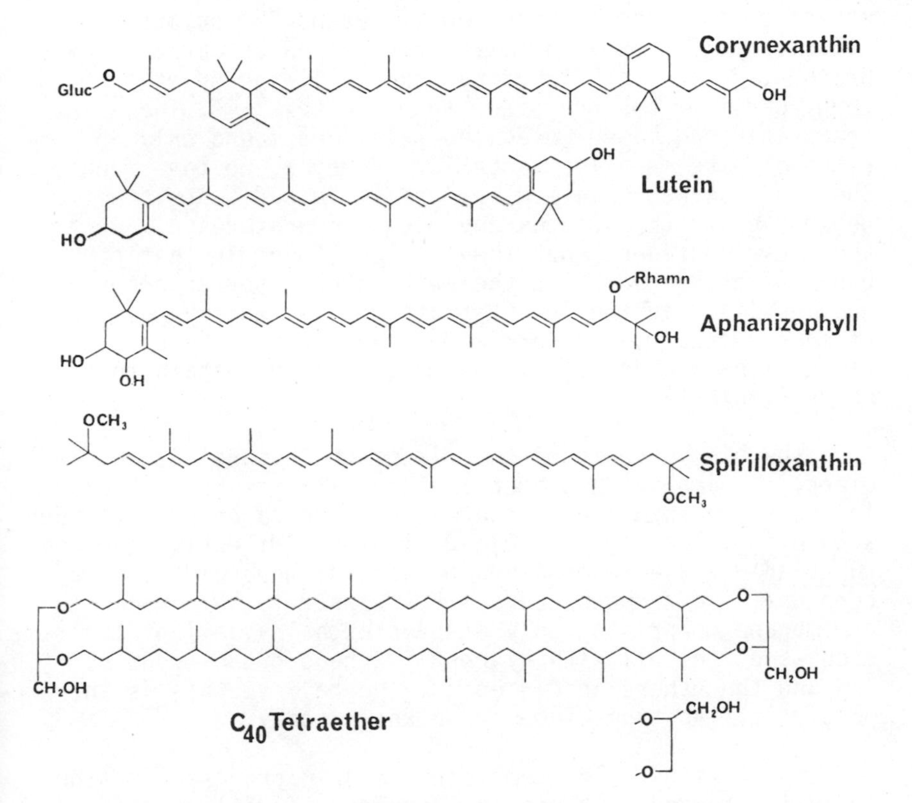

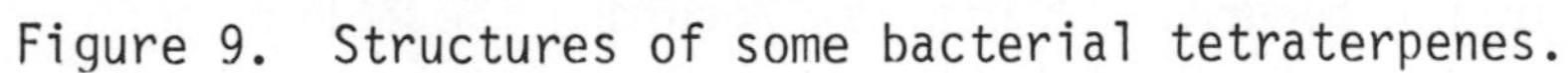

Figure 9. Structures of some bacterial tetraterpenes.

However, it is remarkable that, in the last years, tetraterpenes other than carotenoids have proved to be associated in structures compatible with a role of (nonrigid) membrane bracers, and this, in what is probably the most primitive phylum of micro-organisms known today: the archaebacteria[22]. Here again, knowledge is scanty. However, one can summarize its present state by saying that (all, or some?) archaebacteria have lipid membranes made of glyceryl ethers of the acyclic diterpene, phytanol[23], or its near relatives. The only additives recognized so far are obvious bracers, comprising diphytanyl chains linked by ether bonds to glycerol or, on one side, to a nonitol[24,25]. These remarkable C_{40} based molecules have been found only in extremely thermophilic bacteria. However, we have found their unmistakable molecular fossils in sediments, the geology of which excludes any high temperatures[26]. We are therefore confident that they will be found in that other group of archaebacteria, the methanogens, strict anaerobes now localized in the lower strata of sedimentary muds, but of course potentially more widespread in pre-photosynthetic times. The archaebacteria we have studied contain no hopanoids (Table 1).

The biosynthesis of the dimers of phytanol is not clear. It implies a symmetrical coupling of geranylgeraniol in the sense opposite to that leading to carotenoids. Such a coupling, leading to a C_{40}-diol, would directly give the function we have recognized in sterols, hopanoids and carotenoids. The archaebacteria thus appear to have made use of a membrane comprising only isoprenic chains, all of the same cross-section, all with hydrophilic head-groups, but some in C_{20} and the others in C_{40} units. We believe this is the most primitive type of membrane known today.

Conclusions. The complete scenario proposed for the evolution towards sterols is therefore the following:

(a) - Establishment of the terpene biosynthetic pathway. This leads to geraniol, farnesol, geranylgeraniol and polyprenols.

(b) - From geranylgeraniol, maybe as a glycerol ether, the production of the first membrane, not stabilized.

(c) - From geranylgeraniol, formation of the α,ω-diol by coupling and reduction and of phytanol - (and their glyceryl ethers) and production of the first stabilized membrane, with bracers.

(d) - From geranylgeraniol, the formation of phytoene

(e) - Formation of carotenoid diols or diglucosides, and displacement of the nonrigid di-phytanyl bracers by rigid carotenoid ones.

(f) - "Minor" changes on the carotenoids: C_{50} and C_{30}, squalene being produced as the precursor of the latter.

(g) - Cyclization of squalene to hopanoids. First membranes with really rigid inserts.

(h) - Oxidation of squalene and cyclization of squalene epoxide to 3-hydroxyhopanoids.

(i) - Mutation of the squalene-oxido cyclase, leading to lanosterol, and thence to sterols.

Steps (a) to (g) are compatible with pre-photosynthetic conditions. Step (d) is required for the onset of photosynthesis as we know it, where carotenoids are always involved.

We had stated that our review would start with a consideration of prebiotic systems. This is certainly premature. The only comment which can be made so far concerns the probability of self-organization of appropriate membrane components[27]. A combination of largely hydrophilic proteins, capable of forming lipophilic trunks about 40 Å long, of fluid amphiphilic molecules about 20 Å long, and of one of the reinforcing partners mentioned above (either 20 Å long, rigid and amphiphilic, or 40 Å long, amphiphilic and rigid, or bipolar and rigid or not), any such combination of properly dimensioned and functionalized molecules would certainly organize itself into a system which would be to all intents and purposes a prototype membrane.

MICROEVOLUTION OF TRITERPENES AND STEROLS

Triterpenoid Variation. In the preceding part of the chapter, only a very few essential triterpenoid substances have been mentioned: two C_{30} triterpene monoalcohols, diplopterol and tetrahymanol, and two more, lanosterol and cycloartenol, as biosynthetic precursors of sterols.

Yet, hundreds of triterpenes are known in nature. Obviously, variability of structures is incompatible with a common function. Indeed, no clear biochemical function is known for triterpenoids. Similarly, no clear biochemical function is known for the hundreds of steroid derivatives (sapogenins, heart poisons, alkaloids, insect hormones) present in plants. In a few cases, definite ecological roles are known, either protective (bitter or toxic substances, etc.), or otherwise (feeding attractants, for instance).

We shall assume that this variability is the result of evolutionary drift, and of selection by biochemical, physiological and ecological constraints. In some cases, the evolutionary drift is more limited. We shall briefly review one such case, that of the genus Euphorbia, because some of our conclusions fit nicely with the hypothetical scheme put forward for the hopanoid-sterol change.

Euphorbia Triterpenoids. Euphorbia is a huge, polymorphic genus, containing at least 1500 species almost all of which are lactiferous. These comprise temperate and tropical herbs and shrubs, annuals and biennials, most as prostrate succulents, or spherical xerothermophilic forms; some species are thorny, but their thorns are of diverse morphogenetic origin; among the thorny, cactus-like, typically African, species, some are almost trees while others are dwarf forms; finally there are several groups which show a coral-like morphology. Leaves can be permanent, ephemerous, or absent. In brief, few plant genera display such an extraordinary diversity, which is only very partially illustrated in the Figures shown.

Scattered analyses[28] of the constituents of the latex of several species of Euphorbia showed that many triterpenes are present, as well as less innocuous irritant diterpene esters of the phorbol family (Figure 10). In fact, we showed that some order could be discerned in the 80 species

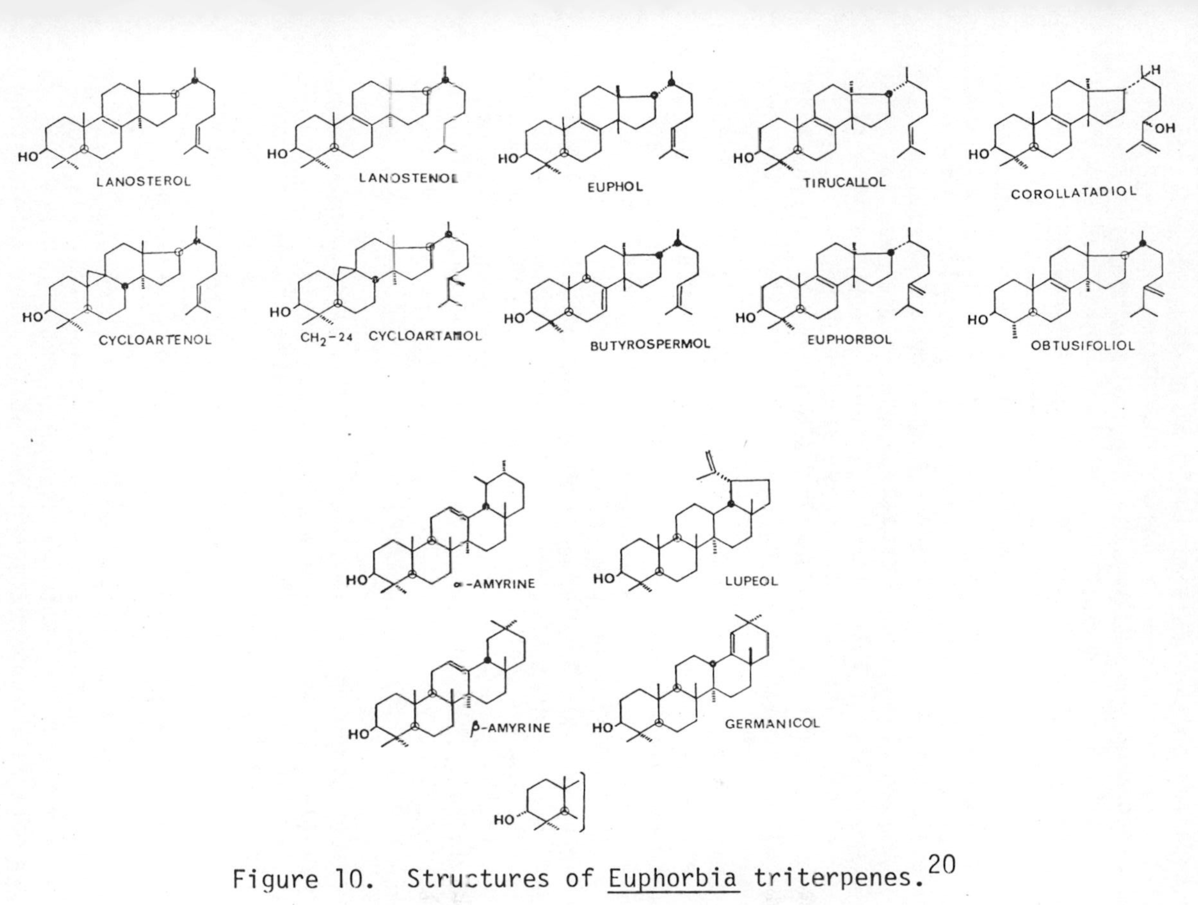

Figure 10. Structures of <u>Euphorbia</u> triterpenes.[20]

studied[28]: the triterpene composition of the latex was more or less constant in the various sub-groups of the genus recognized on the basis of their morphology (Figures 11-14, Table 2). We have fully confirmed, and refined, our previous conclusions, on the basis of much more extensive surveys by Mukam[29] and by one of us[20], covering now about 300 species (about 20% of the genus). We present here only our main conclusions.

It is usually accepted that, in any plant taxon, the most primitive species are those less specialized, not confined to extreme ecological niches; they are normally shrubs. Herbaceous and tree-like species, as well as those occupying highly specialized habitats, are probably derived from these. The most primitive groups of *Euphorbia* species from this point of view are the sub-genus *Esula* (syn. *Tithymalus*) on the Eurasian continent, and the American *Chamaesyce*, now becoming pan-tropical. Both are characterized essentially by the presence in their latex of about 50% or more of cycloartenol and 24-methylenecycloartanol. The same triterpenes are present in several other groups of species, though often as secondary components. However, they are absent, or minor, in some groups which are obviously more advanced as well in many of the typical African cactus-like xerothermophilic groups which are morphologically quite specialized (which contain mostly euphol and euphorbol), and also in the typical American subgenus (or genus) *Poinsettia*. The latter is quite segregated in that the latex is devoid of free tetracyclic triterpenes and contains instead only esters of pentacyclic triterpenes.

Another major variant is found among the coral-like *Tirucalla*: those growing on Madagascar contain in particular tirucallol in their latex, like a few of the South African species. In contrast, the other African coral-like species, which are distinguishable from *Tirucalla* by their pollen morphology and are found as far north as the Arabian Peninsula, do not contain tirucallol but cycloartenol.

The geographic origin of *Euphorbia* is considered to be, before their tectonic separation, South Africa/Madagascar[30]. We interpret our results by assuming that primitive *Euphorbia* latex was characterized only by the accumulation of the triterpenes *normally* produced in every green plant, i.e. of cycloartenol and 24-methylenecycloartanol. We

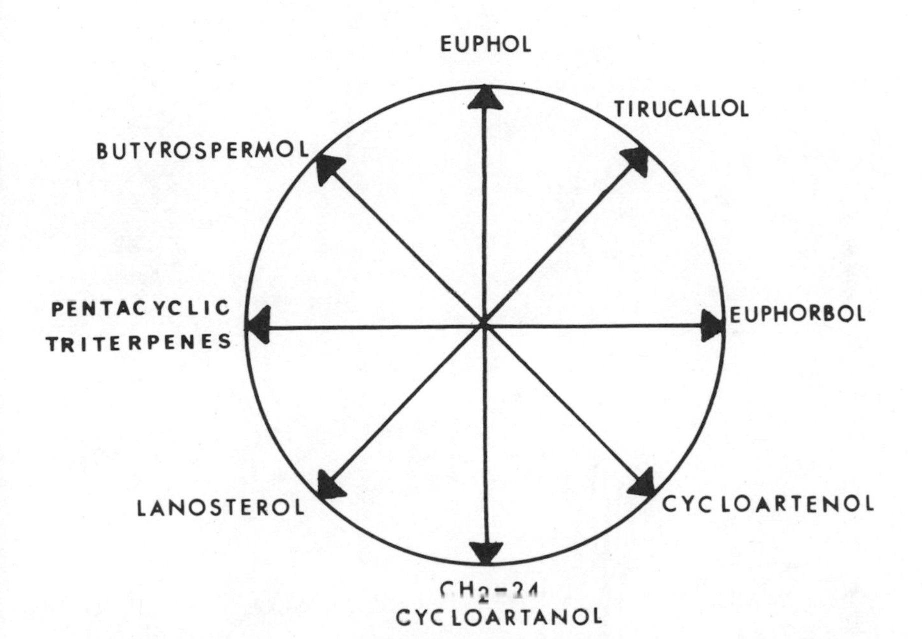

Figure 11. Euphorbia latex types. The relative proportions of the various triterpenes is indicated by thick bars along the diameters of the circle. Type A: mostly euphol and/or euphorbol or tirucallol. Type B: Euphol and tirucallol, without euphorbol. Type C: Mostly cycloartanol and 24-methylenecycloartanol. Type D: Euphol, lupeol, cycloartanol, and 24-methylenecycloartanol, 1:1:1:1. Type E: Esters of pentacyclic triterpenes.

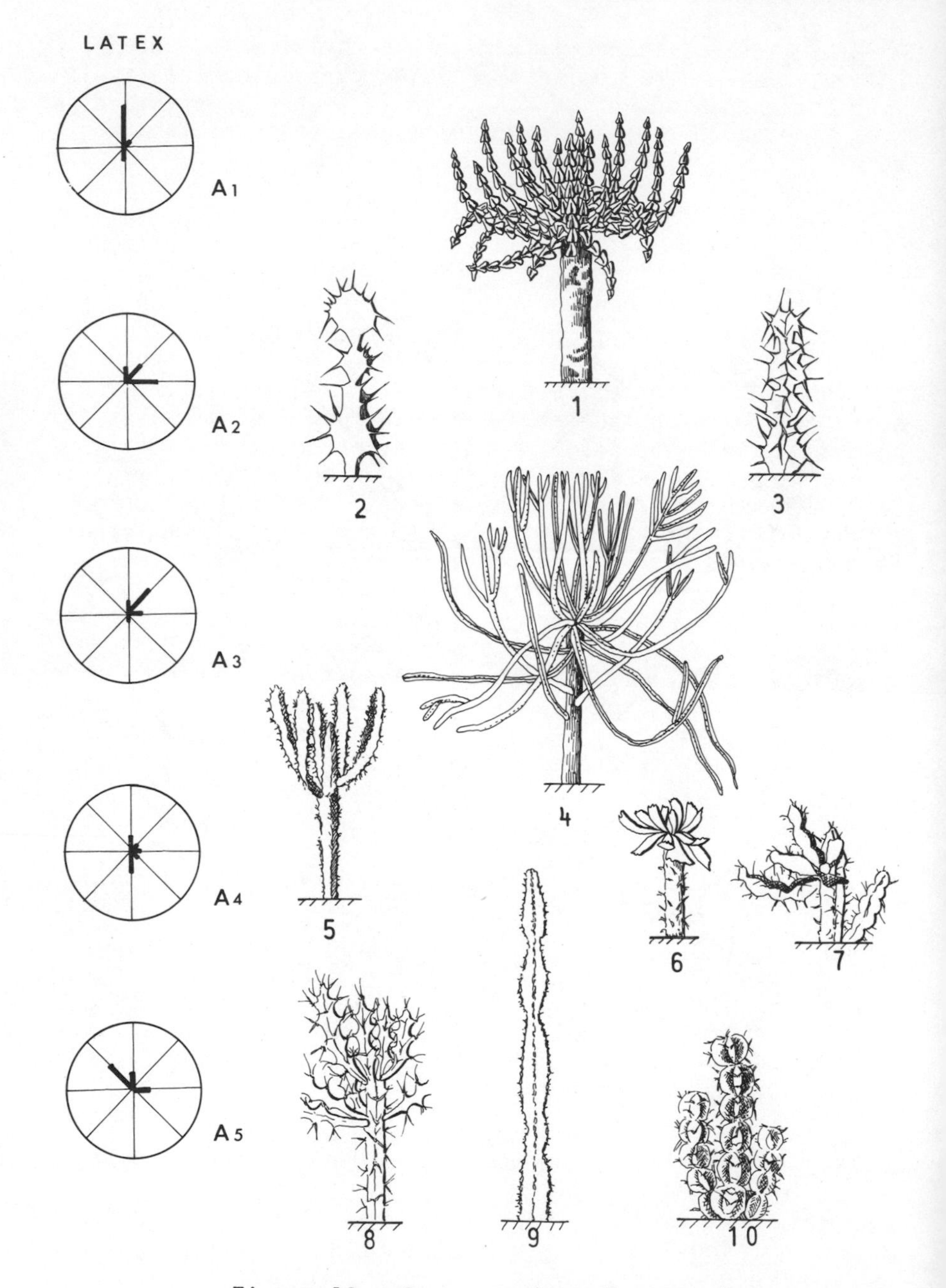

Figure 12. Some species of group A.

LATEX

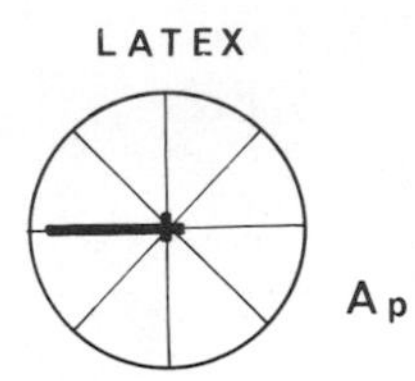

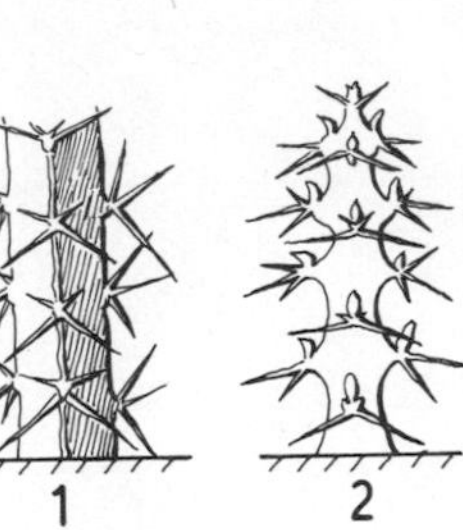

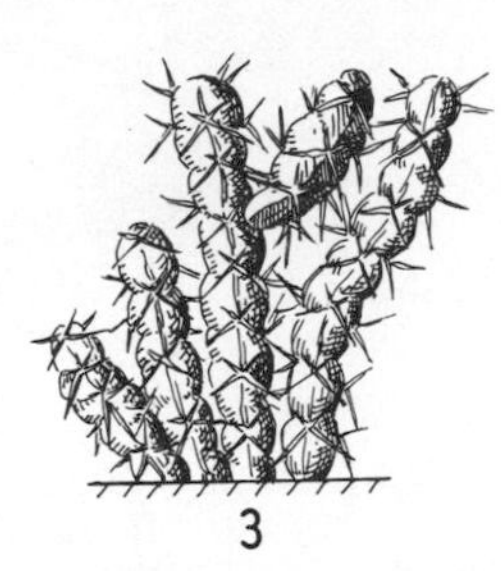

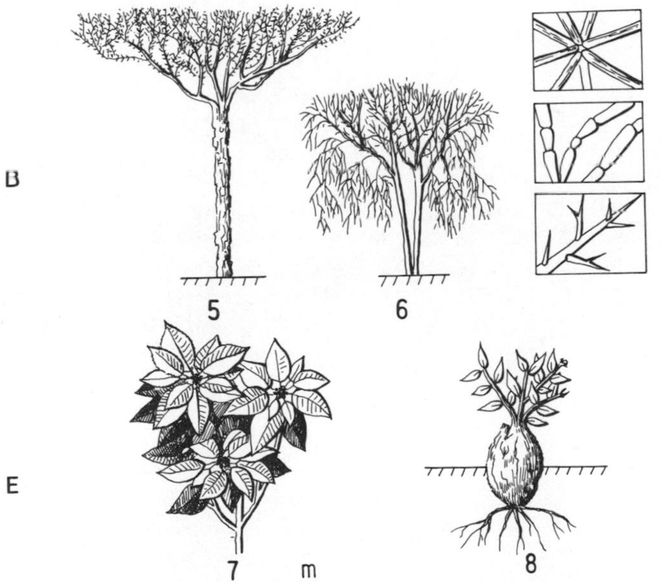

Figure 13. Some species of groups A, B, D, and E.

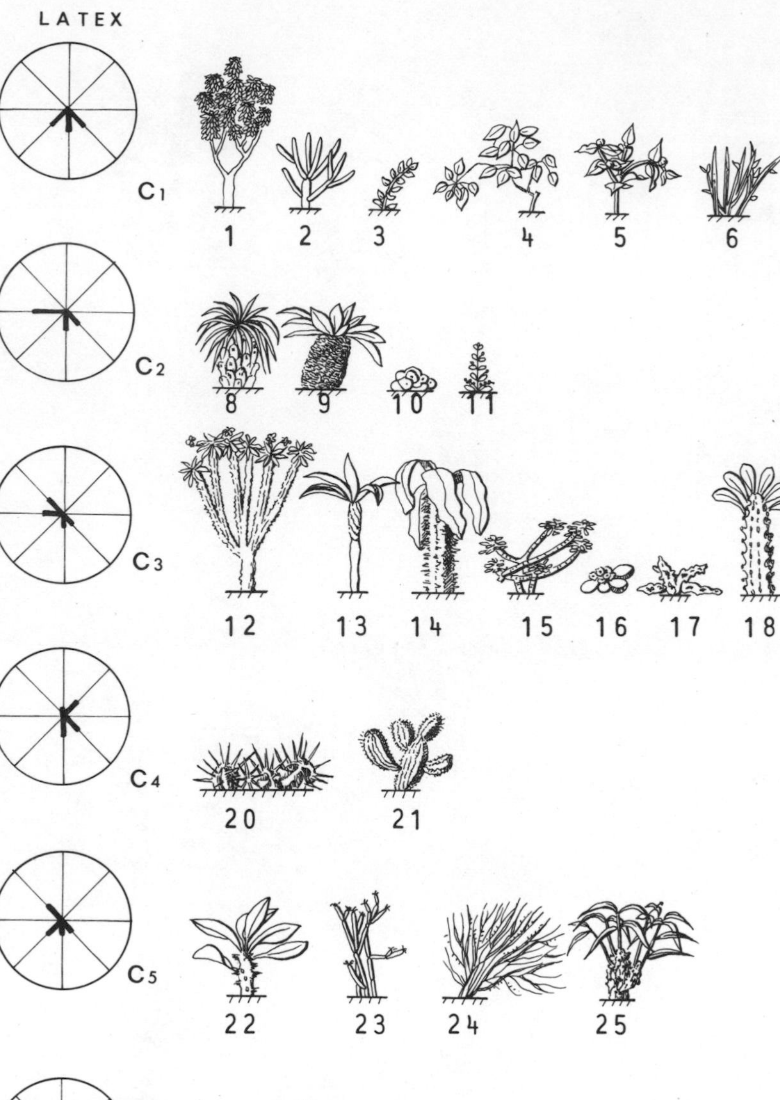

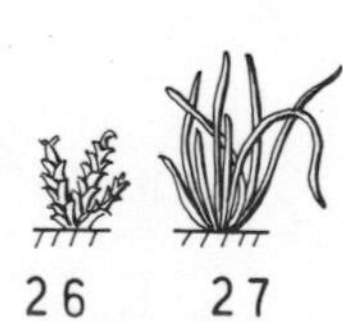

Figure 14. Some species of group C.

Table 2. *Euphorbia* species illustrated in Figs. 12-14.

Figure	No.	Species
12	1.	*E. cooperi*
	2.	*E. angularis*
	3.	*E. grandicornis*
	4.	*E. gracilicaulis*
	5	*E. pseudotriangularis*
	6.	*E. sudanica*
	7.	*E. resinifera*
	8.	*E. coerulescens*
	9.	*E. ammak*
	10.	*E. cooperi*
13	1.	*E. subsalsa* var. Kaokensis
	2.	*E. sp.* (*Tetracanthae*)
	3.	*E. subsalsa*
	4.	*E. beaumeriana*
	5.	*E. laro*
	6.	*E. tirucalli*
	A.	*E. tirucalli* (detail)
	B.	*E. enterophora* (detail)
	C.	*E. stenoclada* (detail)
	7.	*E. pulcherima*
	8.	*E. trichadenia*
14	1.	*E. balsamifera*
	2.	*E. aphylla*
	3.	*E. myrsinites*
14	4.	*E. cotinifolia*
	5.	*E. lathyris*
	6.	*E. pteroneura*
	7.	*E. helioscopia*
	8.	*E. bupleurifolia*
	9.	*E. clava*
	10.	*E. tridentata*
	11.	*Chamaesyce pilulifera*
	12.	*E. soaniarensis*
	13.	*E. ankarensis*
	14.	*E. vignieri*
	15.	*E. francoisii*
	16.	*E. quartziticola*
	17.	*E. cylindrifolia*
	18.	*E. royleana*
	19.	*E. cyparissias*
	20.	*E. horrida*
	21.	*E. mammillaris*
	22.	*E. lophogona*
	23.	*E. sipolisii*
	24.	*E. mauritanica*
	25.	*E. monteiroi*
	26.	*E. hamata*
	27.	*E. antisyphilitica*

assume therefore that all the other variations of latex composition are typical examples of non-Darwinian evolution. For instance, the ancestor of all the cactus-like African Diacanthium, etc. is assumed to have undergone a mutation towards the production of euphol and euphorbol with evolutionary neutral results for the plant since the function of these triterpenes merely is, if any, to be present and constitute the latex.

An extreme case of non-Darwinian evolution of the polyterpene metabolism by a normal mode of biosynthesis running amok, would be the formation of gutta-percha, or rubber: the latex of Hevea, another of the Euphorbiaceae, can be considered as the monstrous depository of the products of uncontrolled condensation of monoterpene units, although we have also found it to contain some cycloartenol[28].

Poinsettia, which occurs farthest away from the region of origin of Euphorbia, are also quite different from other species in that their latex (containing esters of pentacyclic triterpene) is entirely dissimilar from the others. The Chamaesyce, also on the American continent, may be transition forms since besides cycloartenol, they also contain esters of pentacyclic triterpenes.

The most important other general finding in this genus has been that, quite often, within an homogeneous group of species, one is found to contain, beside the usual latex constituents of that group, one or two novel compounds. For instance, in one new species from South-West Africa, hardly distinguishable from E. mauritanica or E. nubica (coral-like, African, with cycloartenol and 24-methylenecycloartanol dominant), we have found also about 50% of α-amyrin. This is not an isolated instance[20].

It appears therefore that, in Euphorbia latex, mutations occur quite easily. Most of the latices contain mixtures of triterpenes, implying that the squalene-oxido cyclases they contain are diverse. The remarks made above concerning the possibility of explaining the transformation from one type of triterpene to another by very small changes in the relative position of functional sites of essentially the same enzyme apply here. We illustrate them for the formation, from similar enzyme-bound intermediates, of either tetracyclic triterpenes (like euphol or tirucallol), or of

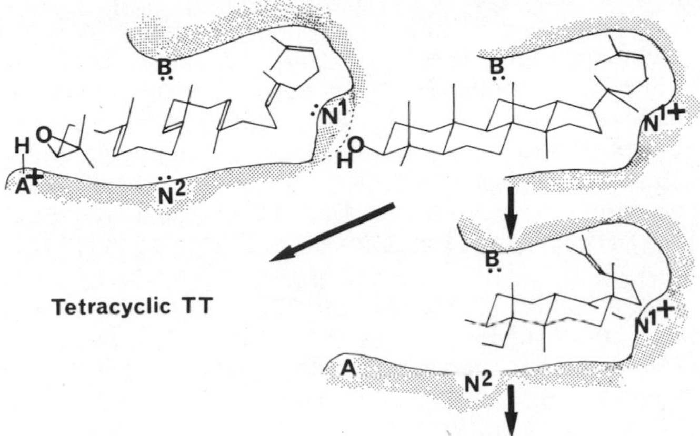

Figure 15. Schematic representation of the enzymatic biosynthesis of tetracyclic triterpenes (euphol series) and pentacyclic triterpenes.

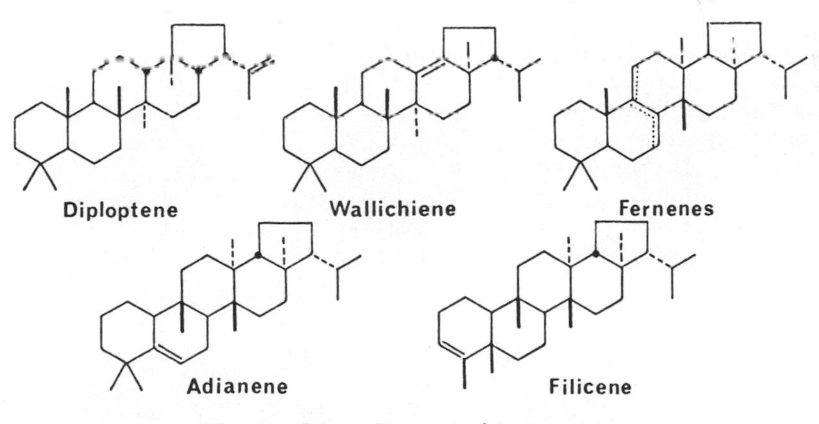

Figure 16. Fern triterpenes.

pentacyclic triterpenes (Figure 15). In every case, rearrangement occurs (this is the common feature of all *Euphorbia* triterpenes), but its extent is probably controlled by the exact positioning of the second nucleophilic group N^2:.

We thus see here one illuminating example of the occurrence of the same mechanisms we have already postulated for the macroevolutionary steps from hopanoids to sterol precursors, but operating with no profound consequences at the microevolutionary level because in the particular tissues considered variations are non-Darwinian, and inconsequential: the triterpenes are there only *end products*.

To return briefly to hopanoids, the remarkable series of 3-desoxy rearranged triterpenes isolated from ferns[31] (Figure 16) (and some other nonvascular cryptogams) can be understood only, we believe, if they are produced by enzymes progressively modified by moving N^2: stepwise nearer the initiating center A-H. If they had any precise biochemical function, such a process would have been impossible: they must therefore be non-functional and be open to non-Darwinian evolution.

ACKNOWLEDGEMENTS

One of us (GO) acknowledges gratefully the help provided by many eukaryotes: by the anonymous person who provoked him to try to integrate experimental results into a more conceptual framework, by Prof. Schofeniels (Liege), who patiently listened to a first version of these ideas, and by many who participated in discussions on some of the points raised. We are grateful to the Centre National de la Recherche Scientifique for continuous support to our laboratory.

REFERENCES

1. Demel, R. A. and B. De Kruyff. 1976. Biochim. Biophys. Acta, 457:109.
2. Asselineau, J. 1962. Les Lipides Bacteriens, Hermann Pub., Paris, pp. 162 and 287.
3. Kushwaha, S. J., E. L. Pugh, J. K. G. Kramer and M. Kates. 1972. Biochem. Biophys. Acta, 260:492.
4. Levin, E. Y. and K. Bloch. 1964. Nature 202:90; Suzue, G. K. Tsukada, C. Nakai and S. Tanaka. 1968. Arch. Biochem. Biophys. 123:644.
5. Bouvier, P. 1978. Dr. Sc. Thesis, Universite Louis Pasteur, Strasbourg.
6. Razin, S. and S. Rottem. 1978. Trends Biochem. Sc., March:51.
7. Bird, C. W., J. M. Lynch, F. J. Pirt and W. W. Reid. 1971. Tetrahedron Lett.:3189.
8. Ourisson, G., P. Albrecht and G. Ourisson. 1978. Pure Appl. Chem. in press.
9. Rohmer, M. 1975. Dr. Sc. Thesis, Universite Louis Pasteur, Strasbourg.
10. Rohmer, M., P. Bouvier and G. Ourisson. 1978. In preparation.
11. Rohmer M. and G. Ourisson. 1976. Tetrahedron Lett.: 3637.
12. Conner, R. L., J. R. Landrey, C. H. Barns and F. B. Mallory. 1968. J. Protozool. 15:600.
13. Barton, D. H. R. and G. P. Moss. 1966. Chem. Commun.:261.
14. Zander, J. M., J. B. Greig and E. Caspi. 1970. J. Biol. Chem. 245:1247.
15. Towe, K. M. 1978. Nature 274:657.
16. Eschenmoser, A., L. Ruzicka, O. Jeger and D. Arigoni. 1955. Helv. Chim. Acta, 38:1890.
17. Anding, C., M. Rohmer and G. Ourisson. 1976. J. Am. Chem. Soc. 98:1274.
18. Barton, D. H. R., T. R. Jarman, K. C. Watson, D. A. Widdowson, R. B. Boar and K. J. Damps. 1975. J. Chem. Soc., Perkin Trans. I:1134.
19. Anding, C. 1973. Dr. Sc. Thesis, Universite Louis Pasteur, Strasbourg.
20. Anton, R. 1974. Dr. Sc. Thesis, Universite Louis Pasteur, Strasbourg.
21. Rees, H. H. , H. J. Goad and T. W. Goodwin. 1968. Biochem. J. 107:417.

22. Woese, C. R., and G. E. Fox. 1977. Proc. Natl. Acad. Sci. USA 74:5488.
23. Kates, M. 1972. in Ether Lipids, Chemistry and Biology. F. S. Snyder, Ed. Academic Press, New York, p. 351.
24. De Rosa, M., S. De Rosa, A. Gambacorta and J. Bu'Lock. 1977. J. Chem. Sc. Chem. Commun. 514; De Rosa, M., S. De Rosa, A. Gambacorta, L. Minale and J. Bu'Lock. 1977. Phytochemistry 16:1961.
25. Langworthy, T. A. and W. R. Mayberry. 1976. Soc. Gen. Microbiol. Proc. 3:165.
26. Michaelis, W. and P. Albrecht. 1975. Unpublished data.
27. Tanford, C. 1978. Science 200:1012.
28. Ponsinet, G. and G. Ourisson. 1968. Phytochemistry 7:89; 1968. Adansonia 8:227.
29. Mukam, L. 1970. Dr. 3e Cycle, Universite de Strasbourg.
30. Croizat, L. 1964. Candollea 19:17; 1965. Webbia 20:573; 1967. ibid. 22:83; 1972. ibid. 27:1.
31. Berti, G. and F. Bottari. 1969. Progr. Phytochem. 1:589.

Chapter Six

REGULATION OF TERPENOID BIOSYNTHESIS IN HIGHER PLANTS

CHARLES A. WEST, MARK W. DUDLEY AND MICHAEL T. DUEBER

Division of Biochemistry
Department of Chemistry
University of California
Los Angeles, CA

INTRODUCTION

Consider the following general facts in relation to the need for regulation of the pathways for biosynthesis of terpenoid compounds in higher plants. Higher plants produce a wide array of terpenoid compounds including monoterpenes, sesquiterpenes, diterpenes, triterpenes and sterols, tetraterpenoid carotenes and xanthophylls, long-chain polyprenyls, and mixed terpenoids that contain an isoprenoid moiety as an integral part of their structures. A single plant species may produce many terpenoid substances including representatives from most or all of these groups. Some terpenoids, such as sterols, gibberellins, carotenes, and the chlorophyll pigments with their polyprenyl side-chains, presumably are produced in all higher green plants. Other terpenoids are formed much more selectively by a few species. Some terpenoids are produced in large quantities and others only in traces.

Figure 1 serves as a reminder that the terpenoid end-products can be considered as products of pathways branching from intermediates of a common central pathway. This representation of all classes of end-products emanating from common pools of intermediates is doubtless an oversimplification for most systems and hence incorrect. Even so, competition by more than one enzyme for a branch-point metabolite is undoubtedly a frequent occurrence in terpenoid biosynthesis. Regulation typically occurs at such points to govern the partitioning of the branch-point metabolite between its alternative fates.

In some instances terpenoids play the role of highly specialized regulatory agents in higher plants. For example, gibberellins and abscissic acid serve as plant hormones that help to regulate various aspects of growth and development, and, as will be discussed later, certain sesquiterpenes and diterpenes act as agents in the defense of plants against invasion by potentially pathogens. In keeping with such functional roles, it is not surprising to find that the production of these terpenoid substances by the plant can be modified in response to environmental stimuli.

All the above factors seem to argue strongly for the existance in plants of regulatory mechanisms to control the production of the array of terpenoid substances. The genetic make-up of the plant will determine the potential for synthesis of different types of terpenoids, but there must also be regulation of the expression of that potential in terms of both the amounts and the activities of the enzymes involved.

Knowledge of the enzymes involved in terpenoid biosynthetic pathways in higher plants and their regulatory features is still fragmentary. Rather than attempting a survey of what is known about regulation of all areas of terpenoid metabolism, we would like to concentrate in this chapter on some aspects of the regulatory features of diterpene biosynthesis that we have been investigating in our own laboratory. Two higher plant pathways will be considered, one leading to the biosynthesis of the gibberellin hormones and another leading to a diterpenoid stress metabolite called casbene, which is produced in young castor bean (<u>Ricinus communis</u> L.) seedlings. In both instances our

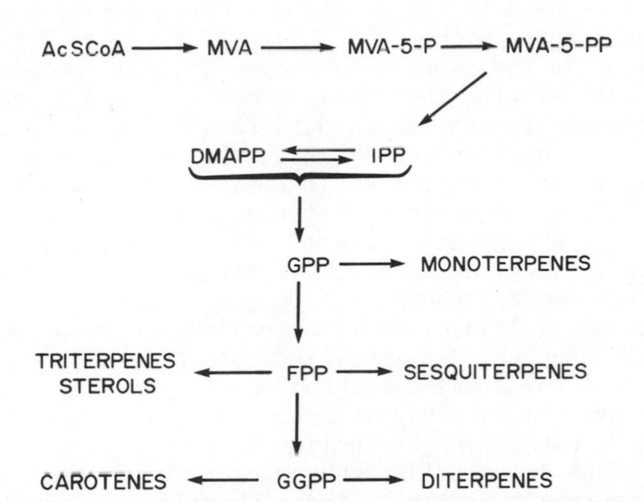

Figure 1. Branching from the general terpenoid biosynthetic pathway. AcSCoA = acetyl-SCoA; MVA = mevalonic acid; IPP = isopentenyl-PP; DMAPP = dimethylallyl-PP; GPP = geranyl-PP; FPP = farnesyl-PP; GGPP = geranylgeranyl-PP.

understanding of the regulatory influences is far from complete, but some observations have been made which point to the existence of regulation and give an indication of the types of mechanisms which may be involved.

REGULATION OF GIBBERELLIN BIOSYNTHESIS

Biosynthetic pathway. Detailed information has been obtained from many laboratories about the biosynthetic pathways leading to the production of gibberellins in cell-free systems from both higher plants and the gibberellin-producing fungus, *Gibberella fujikuroi*.[1] Figure 2 summarizes four stages in the overall pathway. Stage 1 includes the general pathway with the steps common to the biosynthesis of all classes of terpenoids as outlined in Figure 1. Stage 2 is a two-step cyclization in which the acyclic precursor (I) is converted to the tetracyclic diterpene hydrocarbon *ent*-kaurene (II) *via* a bicyclic intermediate not shown. Stage 2 is probably an important site of regulation, as will be discussed later. The enzymes for stages 1 and 2 are soluble rather than membrane-bound and do not require O_2. In contrast, stage 3 involves microsomal membrane-bound oxygenases. It appears that some, and possibly all, of these enzymes of stage 3 are cytochrome P-450-dependent mixed function oxygenases; they all require O_2 and an external source of electrons best supplied by NADPH. The last step in this stage is an oxidative contraction of the B-ring of the *ent*-kaurene skeleton to the gibberellane skeleton (III). The steps in stages 1 through 3 appear to be common to all gibberellin-synthesizing systems examined to date. Stage 4 represents multi-step transformations by largely oxidative reactions of III to a variety of end-product gibberellins. The pathways of stage 4 and the types of C_{19}- and C_{20}--gibberellins formed as major end-products are variable from system to system. The enzymes participating in stage 4 reactions have not been studied in any detail. At least some of them are soluble rather than membrane-bound, and some evidence suggests that the hydroxylation steps in this stage are catalyzed by non-heme-iron requiring mixed function oxygenases.

A number of environmental factors including light, photoperiod and temperature have been shown to influence the amounts of gibberellins extractable from plants and plant

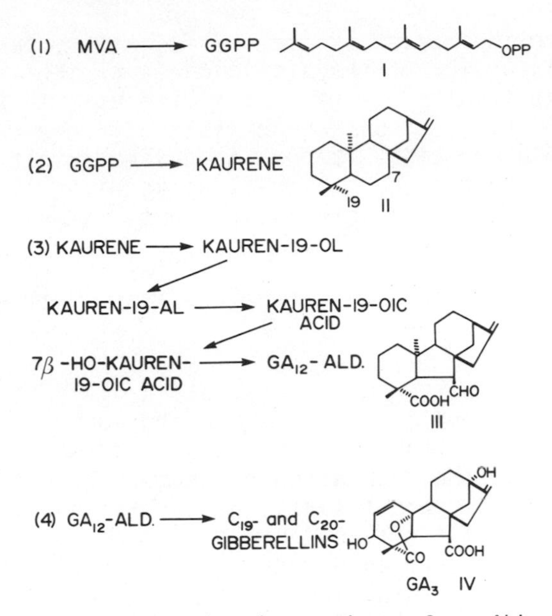

Figure 2. Stages in the pathway for gibberellin biosynthesis. MVA = mevalonic acid; GGPP = trans-geranylgeranyl-PP (I); kaurene = ent-kaur-16-ene (II); 7β-HO-kaurenoic acid = ent-7α-hydroxykaur-16-en-19-oic acid; GA_{12}-ALD = gibberellin A_{12} aldehyde (III); GA_3 = gibberellin A_3 or gibberellin acid (IV).

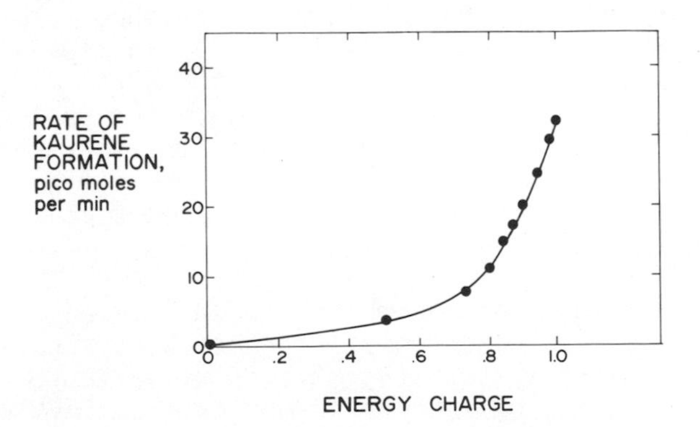

Figure 3. Rates of kaurene synthesis as a function of adenylate energy charge.

parts. (See the review by Jones [2] for references and a critical discussion of the significance of the results.) Although the results are generally difficult to interpret because so many factors can contribute to changes in the levels of gibberellins, it does seem likely that the ability of the plant to produce gibberellins is directly or indirectly responsive to some environmental signals. Thus, one would expect to find regulated steps in the pathway. In the sections which follow, evidence pointing to some possible sites of regulation and regulatory factors will be summarized.

Adenylate energy charge regulation. ATP is recognized as a universal coupling agent for energy-producing and energy-requiring metabolic sequences, and the role of ATP, ADP, and AMP in the regulation of these sequences has been well established, particularly in animals and microorganisms. The concept of adenylate energy charge was developed by Atkinson[3] as a convenient parameter for considering the diverse effects of adenylates in enzyme regulation. The definition of adenylate energy charge is as follows:

$$\text{adenylate energy charge} = \frac{(\text{ATP}) + 1/2(\text{ADP})}{(\text{AMP}) + (\text{ADP}) + (\text{ATP})}$$

The energy charge can vary betweeeen extremes of 0 (only AMP present) and 1 (only ATP present). It is a measure of the fraction of the maximum possible number of phosphoanhydride bonds that are actually present in a mixture of the 5'-adenylates. It has been shown that regulatory enzymes in many biosynthetic (energy-utilizing) sequences show a sharp increase in rate as the energy charge is increased through the physiologically important range of 0.80 to 0.95, and for regulatory enzymes of energy-producing pathways the converse is true[3]. These oppositely directed effects of adenylate energy charge on the activities of energy-utilizing and energy-producing sequences lead to an essential homeostasis of the energy charge *in vivo*.

In order to determine whether adenylate energy charge would have any influence on a terpenoid biosynthetic pathway in a higher plant system, the effects of variation of energy charge on the rate of kaurene formation from mevalonate were tested in crude enzyme preparations derived from the endosperm of immature *Marah macrocarpus* seeds[4]. The striking feature of the response curve (Figure 3)[4] is the sharp increase in the rate of kaurene synthesis at energy charge values above 0.8. This is exactly the type of response predicted for an energy-utilizing biosynthetic sequence. The differences between the responses in the rates of kaurene synthesis to variations in the concentration of ATP alone and to variations in energy charge mixtures can be seen in Figure 4[4]. The upper curve shows the rates of kaurene synthesis from mevalonate at various concentrations of ATP alone, whereas the bottom curve shows the rates in the presence of ATP plus the amounts of AMP and ADP required to give the energy charge values of Figure 3. Clearly either AMP or ADP or both are inhibitory. Additional studies demonstrated that ADP was the inhibitory metabolite while AMP was without effect. The particular step that was affected by variations in energy charge was shown to be the one catalyzed by 5-pyrophosphomevalonate decarboxylase in the general pathway.

These experiments indicate that the rate of kaurene, and hence gibberellin, production may be under energy charge regulation *in vivo*. Biosynthetic sequences in general should be favored when energy is readily available in the cell. The gibberellins are associated with a high rate of vegetative growth in plants, a process which has a high energy requirement. Thus, high rates of production of kaurene and the gibberellin hormones and high rates of vegetative growth may be correlated with energy availability in the plant through energy charge regulation. Of course, the adenylate energy charge is probably but one of a number of regulatory factors that influence the rates of kaurene synthesis from mevalonate. For example, Gray and Kekwick[5] observed the inhibition of mevalonate kinase of *Phaseolus vulgaris* by prenyl pyrophosphates and suggested that this might have physiological significance as a type of feedback inhibition. However, this factor has not been specifically evaluated with respect to kaurene synthesis rates.

Kaurene synthetase as a regulatory site. The overall reaction catalyzed by kaurene synthetase is a two-step cyclization (Figure 5) in which the bicyclic intermediate, copalyl-PP (V), has been shown to participate as an intermediate[6,7]. Step A is formulated as a proton-inhibited cyclization of I followed by the abstraction of a proton to generate the exocyclic methylene function of V. The further transformation of V to kaurene II in step B is thought to

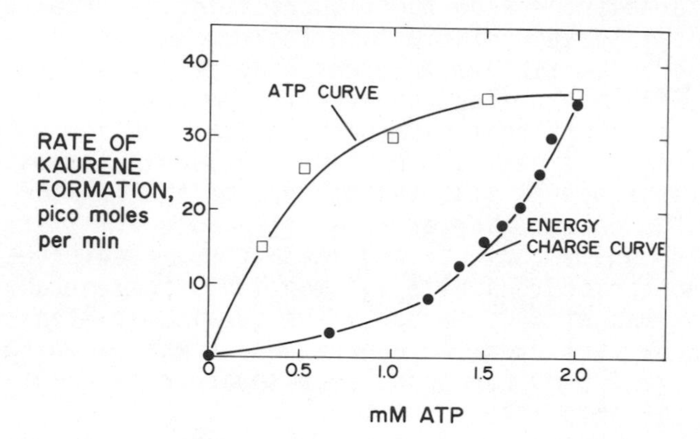

Figure 4. Comparison of the rates of kaurene synthesis as a function of the concentration of ATP alone and ATP in the presence of AMP and ADP.

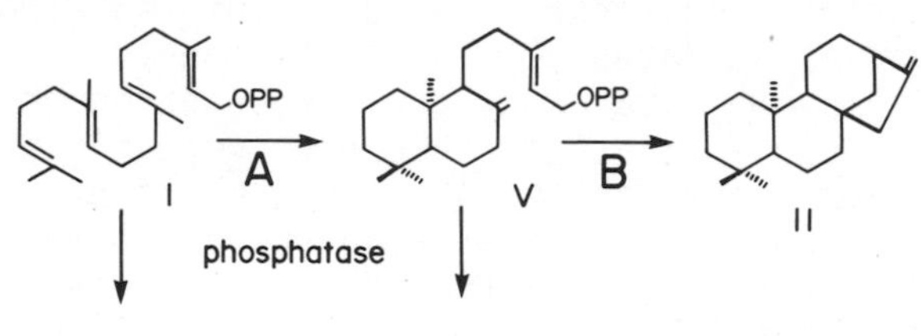

Figure 5. A and B activities of kaurene synthetase. (I) = geranylgeranyl-PP; (V) = copalyl-PP; (II) = kaurene.

involve a complex succession of events including an electronic shift from the exocyclic methylene and elimination of PP_i to form a tricyclic ion (primaradiene skeleton), ring closure by electron migration to a tetracyclic ion (beyerene skeleton), a rearrangement to the kaurene skeleton, and finally abstraction of a proton to generate the exocyclic methylene of the stable product kaurene. The overall reaction catalyzed by the combination of A- and B-activities is referred to as kaurene synthetase AB-activity. AB-activity and B-activity of kaurene synthetase are assayed from the amounts of radioisotopically labeled kaurene formed from labeled geranylgeranyl-PP or labeled copalyl-PP as substrates, respectively. Activity for step A is estimated from the amounts of labeled copalol (after its release from the pyrophosphate ester by phosphatase hydrolysis) plus labeled kaurene formed from radioisotopically labeled geranylgeranyl-PP supplied as the substrate[8,9].

Geranylgeranyl-PP serves as the precursor of several types of terpenoid compounds in addition to kaurene (Figure 6). Carotenoids are formed by the head-to-head dimerization of two molecules of geranylgeranyl-PP[10,11]. Recent evidence indicates that geranylgeranyl esters of chlorophyll are formed from geranylgeranyl-PP and chlorophyllide and then reduced to the corresponding phytyl esters[12,13]. Other mixed terpenoids with C_{20} isoprenoid chains, such as the tocopherols and phylloquinone, probably arise from geranylgeranyl-PP as a precursor as well[14]. In addition, geranylgeranyl-PP is a precursor of both macrocyclic and polycyclic diterpenes other than kaurene[15]. The biosynthesis of carotenes and the phytyl moiety of chlorophyll is known to occur in chloroplasts, and there is increasing evidence that at least a part of the kaurene synthetase activity is localized in chloroplasts, etioplasts and proplastids[16,17]. There is a strong likelihood that all of these reactions involving geranylgeranyl-PP as a substrate occur in the same compartment and, therefore, compete for the same pool of geranylgeranyl-PP. If this is the case, then geranylgeranyl-PP serves as a branch-point metabolite. Regulation of the activities of enzymes utilizing a branch-point metabolite as a substrate is commonly encountered as a means of controlling the partitioning of that metabolite among its various fates in response to cellular requirements. This line of reasoning leads to the suggestion that kaurene synthetase is a logical regulatory enzyme for gibberellin biosynthesis.

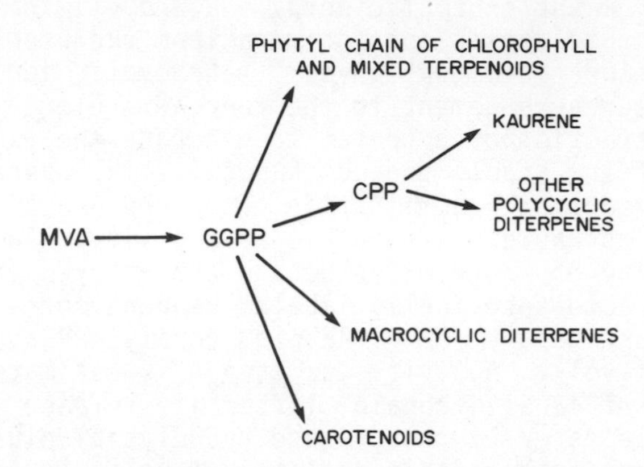

Figure 6. Geranylgeranyl-PP as a branch-point metabolite. MVA = mevalonic acid; GGPP = geranylgeranyl-PP; CPP = copalyl-PP.

Table 1. Characterization of kaurene synthetase preparations

Source	Fold Purification	Type of Kaurene Synthetase Activity	MW	Reference
F. moniliforme mycelia	170	AB	460,000 ± 30,000	19
M. macrocarpus endosperm	10-20	AB	< 45,000	20
R. communis seedlings	15	AB	72,000 - 96,000	8
	190	B_I	76,000 ± 7,000	8
	130	B_{II}	85,200 ± 5,500	8
Spinacia oleracea chloroplasts	13	B	63,000 ± 7,000	8
Pisum sativum etioplasts	--	B	44,000 ± 3,000	8
chloroplasts	--	B	66,000 ± 5,000	8

It should be noted that Maudinas et al.[18] have reported that the formation and utilization of geranylgeranyl-PP for phytoene biosynthesis from isopentenyl-PP in tomato plastids occurs within an enzyme complex. To the extent that there is channeling of geranylgeranyl-PP to specific end-products in such complexes, the above view of geranylgeranyl-PP as a branch-point metabolite would, of course, need to be modified.

Because of the possibility that kaurene synthetase may serve as a regulatory enzyme, there has been an interest in the purification of this enzyme from various sources so that its properties could be examined for indications of regulatory features. A combination of standard fractionation procedures including ammonium sulfate precipitation and anion exchange, gel filtration, and hydroxylapatite chromatography have been employed with soluble extracts of tissue homogenates or organelle lysates. The kinetic properties of the purified enzymes are generally similar. Normal Michaelis-Menten kinetic patterns with Km values in the range of 0.5 to 2 μM are seen with both substrates. A divalent metal ion is required with Mg^{2+} preferred and Mn^{2+} generally less effective. Sulfhydryl reagents are inhibitory to both A-and B-activities, with the B-activity the more sensitive. With the most thoroughly characterized enzymes from the fungus F. moniliforme[8] and the endosperm of immature M. macrocarpus[9] seed, endogenous copalyl-PP generated from geranylgeranyl-PP was used in preference to added copalyl-PP for the formation of kaurene. It was concluded that copalyl-PP might not accumulate to a significant extent under normal physiological conditions because of this channeling effect. Purification of the A- and B-activities of kaurene synthetase from F. moniliforme produced in a constant ratio of specific activities and the mixture could not be resolved[8]. Those from M. macrocarpus were less stable and a constant ratio was not maintained during purification; however, they were not resolved from one another by any of the procedures utilized[9].

Table 1 lists the types of catalytic activities and approximate molecular weights found for some partially purified kaurene synthetase preparations from various sources. The molecular weights of the higher plant enzymes, which were estimated from their elution volumes from a gel filtration column calibrated with reference globular proteins, were in the range of about 40,000 to 100,000 and thus differed significantly from values for the fungal enzyme. The most surprising result from this survey was the discovery of kaurene synthetase preparations from some sources with readily detectable B-activity and little or no A-activity. In the case of *Ricinus communis* seedlings, two forms of B-activity and a heterogeneous AB-activity were resolved from one another by chromatography on DEAE-Sephadex A-25[21]. It is noteworthy that these B-activities were not very different in apparent molecular weight from the AB-activity from the same source. Yafin and Shechter also found only B-activities associated with kaurene synthetase preparations from tissue cultures of *Nicotiana tabacum* and *Lycopersicum esculentum*, but found AB-activity in germinating seedlings of *L. esculentum*[22]. The preparation of an A-activity free from B-activity has not yet been reported.

The origins and significance of these kaurene synthetase preparations with only B-activity are still unclear. So far as is known, the absence of A-activity would result in a lack of kaurene synthetic capacity since there is no other known source of copalyl-PP. It is possible these B-activities may represent nonphysiological artifacts resulting from the selective denaturation, inhibition, or inactivation of A-activity during the course of extraction and purification of the enzyme. However, in the case of the chloroplast and etioplast enzymes it might be noted that the activity pattern was already established in the crude preparation initially assayed. Alternatively, these B-activities may be a consequence of a physiologically important regulation of the A-activity in the tissues from which they are derived. So far, efforts to demonstrate a latent or masked A-activity associated with the B-activity extracted from spinach chloroplasts have not been successful. Treatments with proteolytic enzymes, sulfhydryl agents, and inhibitors of proteolysis were among the approaches attempted. Current work in our laboratory is directed at an understanding of the basis for the absence of significant A-activity in the presence of B-activity that seems to characterize some plant tissues.

The susceptibility of the A-activity to inhibition by certain synthetic plant growth retardants and related substances[7,23] is another feature of kaurene synthetase which has directed attention to the A-activity as a possible site of regulation. The most thorough study of the interaction of these inhibitors with kaurene synthetase of higher plant origin has been with the partially purified enzyme from _M. macrocarpus_ endosperm[9]. Table 2 summarizes the results for a few of the substances tested. Three plant growth retardants plus the steroid synthesis inhibitor SKF 525-A were potent inhibitors of A-activity while displaying little or no effect on the B-activity. Amo-1618 and Phosfon D were noncompetitive inhibitors of the A-activity. This inhibition of kaurene synthesis, and hence of gibberellin synthesis, may be at least partially responsible for the actions of these substances on plant growth. The growth retardants CCC (chlorocholine chloride) and B-995 (_N_, _N_-dimethylaminosuccinamic acid), which had little effect on kaurene synthetase activities at concentrations below 1 mM, must exert their effects on plant growth in some other manner.

The existence of a site for synthetic substances that could regulate the A-activity of kaurene synthetase led to the speculation that there might be natural metabolites which could interact _in vivo_ with this site to regulate the expression of kaurene synthetase activity. A number of substances were tested in the study cited[9], but so far no good candidates for metabolic effectors of kaurene synthetase activity have been discovered. Also, a number of gibberellins were tested as possible feedback inhibitors of kaurene synthetase and found to be ineffective[9]. Therefore, in spite of the logical arguments for the regulation of the A-activity of kaurene synthetase and the indications that it is susceptible to inhibition or inactivation _in vitro_, there is still no evidence of how or whether such regulation may be accomplished _in vivo_.

Regulation of kaurene oxidation by ancymidol. α-Cyclopropyl-α[_p_-methoxyphenyl]-5-pyrimidinemethanol alcohol (Ancymidol or EL-531) (Figure 7) is a synthetic substance which acts as a growth regulator for a wide range of plants by inhibiting internode elongation[24,25]. These growth-retarding effects can be completely overcome by applications of gibberellic acid[24,25]. However, unlike growth retardants such as Amo-1618 mentioned above, ancymidol at concentrations below 0.1 mM had no significant effect on kaurene

Table 2. The effects of plant growth retardants and related substances on the activities of M. macrocarpus kaurene synthetase.[a]

Inhibitor	Approximate concentration for 50% inhibition of:	
	Activity A	Activity B
	μM	μM
SKF-525A [b]	0.5	100-500
Phosfon D [c]	1.0	< 500
Q-64 [d]	1.0	Not inhibitory
Amo 1618 [e]	1.0	Not inhibitory

[a] Data from Frost and West [9]

[b] N,N-Dimethylaminoethyl 2,2-diphenylpentanoate

[c] Tributyl-2,4-dichlorobenzylphosphonium chloride

[d] 2-(N,N-Dimethyl-N-octylammonio)-p-menthan-1-ol bromide

[e] 2'-Isopropyl-4'-(trimethylammonio)-5'-methylphenyl piperidine-1- carboxylate chloride

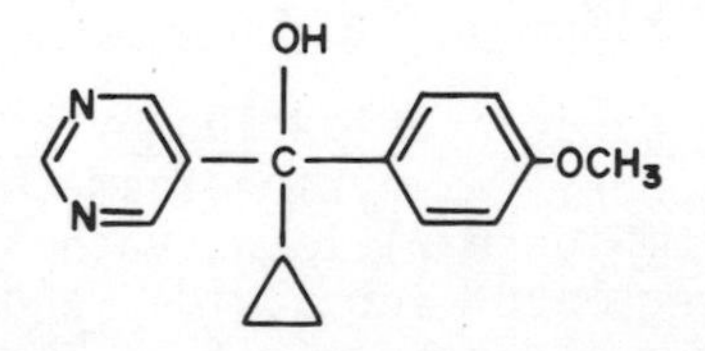

Figure 7. Structure of ancymidol.

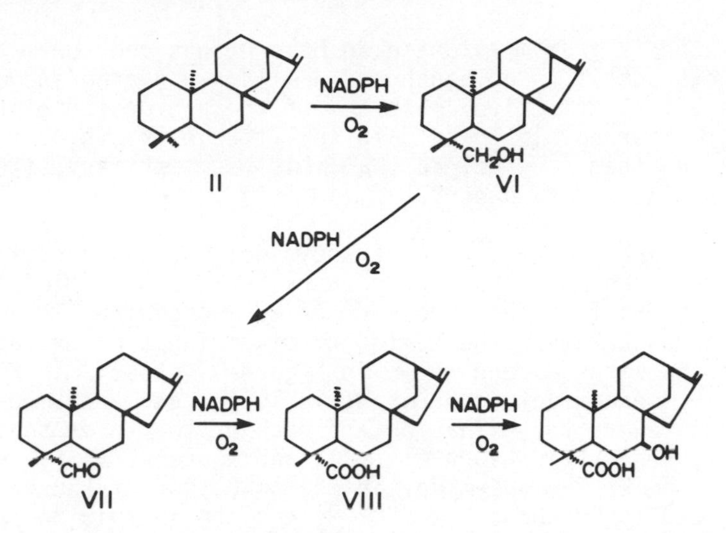

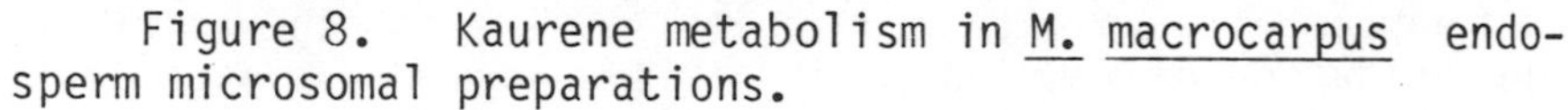

Figure 8. Kaurene metabolism in M. macrocarpus endosperm microsomal preparations.

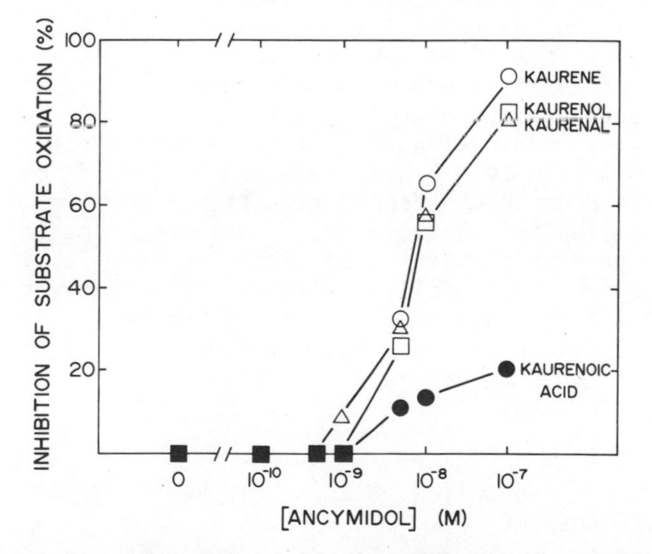

Figure 9. Comparison of the effects of ancymidol on the oxidation of [^{14}C-]kaurene, [^{3}H-]kaurenol, [^{3}H-]kaurenal and [^{3}H-]kaurenoic acid by M. macrocarpus microsomal preparations.

biosynthesis from mevalonate in *M. oreganus* endosperm cell-free extracts[24]. Coolbaugh and Hamilton did find ancymidol to be a very effective inhibitor of the conversion of kaurene to kaurenol in these same cell-free extracts, a result which led them to conclude that this was a primary site of action of the retardant[24].

The inhibitory action of ancymidol for kaurene oxidation has been more thoroughly characterized with microsomal preparations from the endosperm of *M. macrocarpus*[26]. These microsomes catalyze the series of mixed function oxidations beginning with kaurene shown in Figure 8 (stage 3 in Figure 2)[27,28]. Ancymidol is a potent inhibitor and about equally effective for the oxidations of each of the three substrates, kaurene (II), kaurenol (VI), and kaurenal (VII) (Figure 9)[26]. The K_i for ancymidol in the oxidation of kaurene was estimated to be about 2×10^{-9} M, and the inhibition pattern was noncompetitive with respect to kaurene concentration. The oxidation of kaurenoic acid (VIII) was not so significantly affected by ancymidol in this concentration range.

Ancymidol appears to be a very selective inhibitor of the higher plant cytochrome P-450-dependent enzyme system that catalyzes the oxidations of kaurene, kaurenol and kaurenal. It was shown not to give significant inhibitions of the microsomal cytochrome P-450-dependent oxidations of kaurene catalyzed by either animal (rat liver) or fungal (*F. moniliforme*) preparations[26]. It likewise did not affect the rate of cinnamic acid 4-hydroxylation in a different higher plant cytochrome P-450-dependent system from *Sorghum bicolor* that catalyzes this reaction[26]. Thus, ancymidol should prove to be a very useful agent for studies or applications which require the inhibition of gibberellin biosynthesis in higher plants because of its efficacy and selectivity as an inhibitor. There are, however, no known natural agents equivalent to ancymidol in its actions as an inhibitor of kaurene oxidation.

REGULATION OF PRODUCTION OF THE ANTIFUNGAL STRESS METABOLITE CASBENE

Phytoalexins and stress metabolites. Phytoalexins are one class of what have been described more broadly as stress metabolites--substances produced by plants in greatly increased quantitites as a consequence of an environmental or physiological stress imposed on the plant. For the purposes of the discussion here, we will be particularly interested in the regulatory mechanisms which account for this greatly increased capacity of the plant tissues to synthesize these metabolites after exposure to the stress situation. Increasing attention is being given to suspected phytoalexins and the control of their production because of their proposed roles in plant disease resistance[29-33].

Muller and Borger set forth the phytoalexin theory of plant disease resistance in 1940 on the basis of the observed production of antifungal substances by potato plants after they had been exposed to fungi[34]. In commenting on this theory later, Muller defined phytoalexins as "antibiotics which are produced as a result of the interaction of two metabolic systems, host and parasite, and which inhibit the growth of microorganisms pathogenic to plants"[35]. This definition emphasizes two important attributes of phytoalexins. (a) They are not produced, or are produced only in very low levels, in healthy plant tissues, but the interaction of a potentially pathogenic fungus with a plant leads to a greatly stimulated production of the phytoalexins by plant cells in the immediate vicinity of the contact. (b) Phytoalexins are antifungal agents that affect the growth and development of a wide range of fungi. Cruickshank favors the adoption of a more specific definition which not only includes the above elements, but also emphasizes the role of phytoalexins in arresting fungal growth *in vivo* during a hypersensitive response[29,36].

A large number of substances that serve, or have been proposed to serve, as phytoalexins have been identified in higher plant tissues[30]. These are invariably low-molecular-weight substances and most usually have been members of two groups of natural products--isoflavonoids or sesquiterpenes. Related species of the Leguminosae utilize one, or usually several, of a group of biogenetically related isoflavonoids, especially those with a pterocarpan ring system, as phytoalexins. Species of the Solanaceae produce one or more of a group of biogenetically related sequiterpenes for this purpose. Sweet potato (onvolvulaceae) utilizes a different group of sequiterpenes for its defense. This emphasis on isoflavonoids and sesquiterpenes as phytoalexins may be a consequence of the fact the investigations have tended to concentrate on plants of the Leguminosae and the Solanaceae. In a few cases where plants from other families have been examined, different classes of substances have been found as phytoalexins.

Diterpenes as phytoalexins. The first suggestion that a diterpene might function as a phytoalexin resulted from work in our laboratory[37]. Casbene is one of at least five diterpene hydrocarbons which are biosynthesized from either mevalonate or geranylgeranyl-PP in cell-free extracts of young castor bean (Ricinus communis L.) seedlings[15,38]. A tentative structure ([IX] in Figure 10) was proposed for casbene on the basis of chemical and spectral properties of a biosynthesized sample[38]. The properties of a chemically synthesized sample of structure (IX) appear to substantiate the proposed structure for casbene[39]. A postulated pathway to account for the biosynthesis of casbene from geranylgeranyl-PP is illustrated in Figure 10. An intramolecular cyclization is initiated by elimination of PPi from C-1 coupled with an attack of a pair of electrons from the distal double bond to form the macrocyclic carbocation shown; further formation of a cyclopropyl carbonium ion and elimination of a proton from the pro-C-1 position produces casbene (IX). Casbene synthetase has been partially purified and characterized from castor bean seedlings in which casbene synthesis has been elicited[20]. The apparent molecular weight of the enzyme is 53,000. Geranylgeranyl-PP shows normal Michaelis-Menten kinetics (Km = 1.9 μM), and Mg^{2+}most effectively as the required divalent metalion.

In early studies it was seen that the capacity of cell-free extracts of castor bean seedlings for casbene biosynthesis was highly variable. Sitton first recognized a positive correlation between the degree of contamination of the seedlings with fungi during germination and the activity of cell-free extracts for casbene biosynthesis[37]. Rhizopus stolonifer and Aspergillus niger were isolated and identified as the most common fungal contaminants. In controlled experiments it was demonstrated that extracts prepared from seedlings that had received 12-24 hr. exposures in the same Petri dish to agar blocks on which R. stolonifer mycelia were growing gave initial rates of casbene biosynthesis 20-40 times as high as extracts from seedlings maintained under sterile conditions throughout[37]. Exposure to other types of fungi similarly stimulated casbene biosynthesis in cell-free

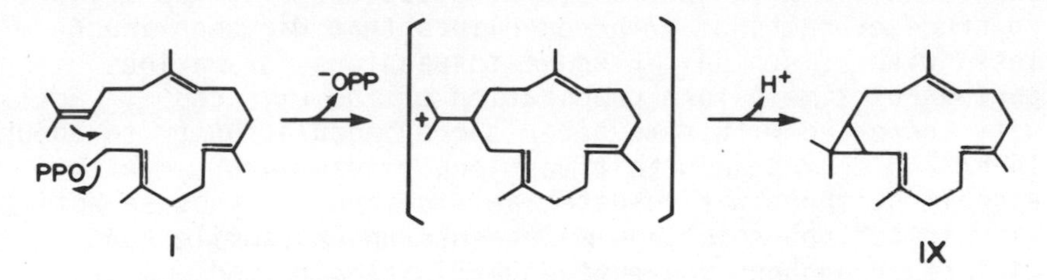

Figure 10. Proposed mechanism for the biosynthesis of casbene (IX) from geranylgeranyl-PP (I).

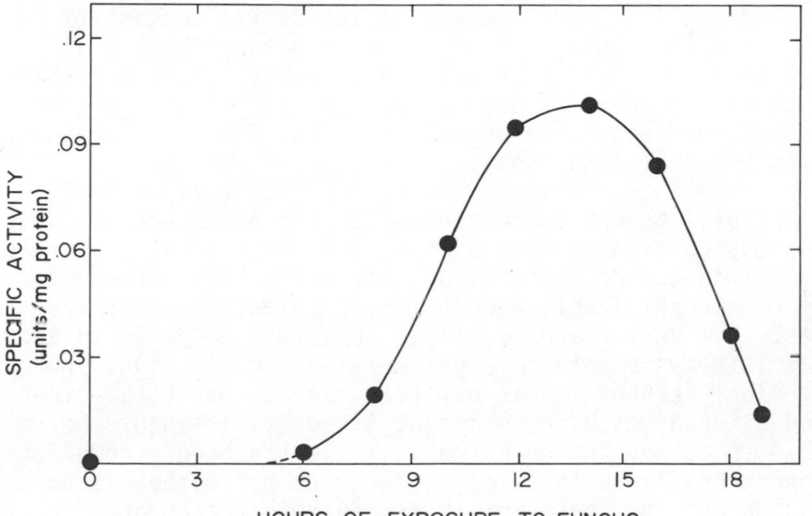

Figure 11. Specific gravity of casbene synthetase as a function of time exposure of castor bean seedlings to Rhizopus stolonifer spores. All seedlings were germinated for 67 hr. At the indicated times before preparation of cell-free extracts for assay of casbene synthetase, the seedlings were inoculated with a suspension of Rs spores in potato-dextrose medium.

extracts. Figure 11 shows the results of a more recent experiment in which casbene synthetase activity was measured in crude extracts of 67-hr seedlings that had been inoculated with _R. stolonifer_ spore suspensions for various periods of time before preparation of the extracts[20]. Activity increased with time after spore inoculation up to about 14 hr and decreased with times longer than 14 hr. In this experiment there was no detectable casbene synthetase activity in control seedlings maintained under sterile conditions throughout the entire germination period.

The antibiotic properties of casbene were tested with an isolated sample prepared by biosynthesis[37]. Ten micrograms per ml in agar was sufficient to give a substantial inhibition in the radial growth of _A. niger_. Even lower concentrations were strongly inhibitory to the growth of _Escherichia coli_ K12 in liquid medium. Similarly 20 μg of casbene applied to the leaves of the _dwarf-5_ mutant of _Zea mays_ inhibited both the endogenous and the gibberellin-stimulated elongation of the leaf sheaths. Although these tests of its activity have been very limited, it appears that casbene acts as a growth inhibitor for a wide range of organisms including fungi.

Thus, casbene possesses two of the important attributes of a phytoalexin: it is biosynthesized in greatly increased amounts in castor bean tissues that have been exposed to any one of several fungi, and it possesses antifungal properties. However, there is as yet no direct evidence to show that it plays a role in disease resistance _in vivo_. For technical reasons it has not been easy to quantitate the production of casbene in intact tissues. It should be possible to do so, but relatively little has been accomplished along these lines to date. Whether or not casbene should be classed as a phytoalexin, it is clear that the potential of castor bean seedlings to produce it is markedly increased after the stress of fungal contamination. Therefore, studies of the regulation of its biosynthesis should have relevance for the general problem of control of stress metabolite production.

The best evidence that diterpenes can serve as phytoalexins comes from the recently published work of Cartwright _et al._[40] which indicates that momilactones A and B act in this capacity in rice (_Oryza sativa_) seedlings. Although specific instances have not been documented to date, it seems reasonable to speculate that other diterpene phytoalexins and metabolites derived from them will be discovered among the large group of diterpenoid substances elaborated by plants. A survey of the literature indicates a surprisingly large number of cases where diterpenoid substances have growth-inhibitory or toxic properties for microorganisms or other organisms or cell-types. Perhaps some of these substances represent antifungal antibiotics which have been produced in the plant in response to challenges by potential pathogens.

Elicitors of casbene biosynthesis. After it was discovered that fungi were responsible for the stimulation of casbene production in extracts of castor bean seedlings, attention was directed to the detection of substances associated with the fungus that are capable of triggering this response during the plant-fungus interaction. Such substances are generally referred to as elicitors following a suggestion of Keen *et al.*[41]. An assay system for casbene elicitors was devised in which the rate of casbene biosynthesis from [2-^{14}C]mevalonate was measured in a cell-free extract prepared from pooled seedlings which had been exposed under sterile conditions to the fraction to be tested[42]. It was possible with this assay to show the presence of elicitor activity in culture filtrates of *R. stolonifer* grown in potato dextrose medium. Sephadex G-25 gel filtration chromatography of the culture filtrate resulted in the separation of high-molecular weight and low-molecular weight fractions with elicitor activity. The low-molecular weight fraction also contained a seedling growth inhibitor and has not been examined further.

The high-molecular weight elicitor fraction has been partially purified by a combination of gel filtration chromatographic procedures[42]. The most active fraction, which was still impure and may have contained more than one active component, was composed of protein and carbohydrate in a ratio of about 2:1. The average molecular weight of the elicitor in 10 mM Na phosphate at pH 7 was estimated at 30,000 ± 5,000 from its elution volume on a calibrated Sephadex G-100 column in comparison with globular protein reference standards. Elicitor activity was completely destroyed by heating at 60°C or 15 min. The activity was also partially destroyed by pronase digestion and completely inactivated by treatment with periodate. From these properties it is concluded that the elicitor is most likely a protein and may be a glycoprotein that depends on both the carbohydrate and protein moieties for its activity.

This casbene elicitor fraction from _R. stolonifer_ differs from phytoalexin elicitors of fungal origin described for other systems[29,42-44] in that it appears to require a native protein structure for activity. Those elicitors which have been sufficiently purified to give meaningful indications of their chemical compositions fall into two general categories: those that are glucan polysaccharides derived from the fungal cell wall and those that appear to contain protein or polypeptide as an integral part of the active elicitor. However, none of them has been fully characterized and there are still unanswered questions about their chemical nature, their modes of action, and their roles in specific disease resistance mechanisms.

Enzymes of casbene biosynthesis. The enzymes most immediately involved in casbene biosynthesis include casbene synthetase and the prenyl transferases which are responsible for the formation of geranylgeranyl-PP from isopentenyl-PP and dimethylallyl-PP. The properties of partially purified casbene synthetase were noted earlier. Two types of prenyl transferases are involved in the biosynthesis of geranylgeranyl-PP. They have been partially purified from castor bean seedlings inoculated with _R. stolonifer_ spores and some of their properties determined. Geranyl transferase (farnesyl-PP synthetase) was purified by a combination of ammonium sulfate fractionation, QAE-Sephadex chromatograph, and Sephadex G-100 gel filtration chromatography[45]. Two isoenzymes (I and II) were largely resolved from one another during the ion exchange chromatography step. There was no evidence of interconversion of the two forms when each was recycled through the ion exchange procedure. Farnesyl transferase (geranylgeranyl-PP synthetase) was purified by a combination of ammonium sulfate fractionation, DEAE-Sephadex A-25 sievorptive chromatography and hydroxylapatite chromatography[19]. After relatively long periods of exposure to the fungus, the sievorptive chromatography step revealed two forms of this enzyme (I and II); however, these were poorly resolved from one another and the relative amounts of each varied as a function of the time of exposure of the seedlings to the fungus. It has been found more recently that only one form of farnesyl transferase activity is seen if the enzyme is extracted into a medium containing relatively high concentrations of sucrose. Thus, we are uncertain at this time about the physiological significance of the two forms of farnesyl transferase.

The properties of these prenyl transferases are summarized in Table 3. There is very little in terms of their properties to distinguish geranyl transferases I and II from one another other than their different behavior on ion exchange chromatography. Similarly, farnesyl transferases I and II have virtually indistinguishable properties except for their pH optima. The differences in the substrate requirements between the geranyl and farnesyl transferases is of considerable interest. The geranyl transferases utilize both dimethylallyl-PP and geranyl-PP quite efficiently as allyl donors and therefore would be expected to synthesize farnesyl-PP readily from C_5 precursors as is generally the case with this enzyme from other sources. However, the farnesyl transferases utilize only farnesyl-PP efficiently as an allyl donor. Thus, under physiological conditions, this enzyme can be expected to catalyze only the C_{15} to C_{20} transformation. This is the first report of a geranylgeranyl-PP synthetase that is in essence an elongation system for the addition of a single prenyl unit. It is concluded that the concerted action of both a geranyl transferase and a farnesyl transferase are required for the synthesis of geranylgeranyl-PP from C_5-isoprenoid units.

Estimates of the number of units of geranyl transferase, farnesyl transferase and casbene synthetase obtained from homogenates of seedlings that had been inoculated with R. stolonifer spores are compared in Table 4 with the levels of those enzymes obtained from homogenates of seedlings that had been maintained under sterile conditions throughout the germination period. All three enzymes were present in significantly higher levels in elicited seedlings. It appears from these data that casbene synthetase is present in elicited seedlings in much higher levels than either of the prenyl transferases. However, it should be pointed out that the estimates of casbene synthetase were made directly in crude extracts whereas it was necessary to remove interfering enzymes from the crude extracts by ion exchange chromatography before estimates of the prenyl transferase activities could be made. Losses of an unknown magnitude may have

Table 3. Properties of prenyl transferases purified from extracts of castor bean seedlings.[a]

Property	Geranyl transferase I	Geranyl transferase II	Farnesyl transferase I	Farnesyl transferase II
Km values (μM)				
IPP[b]	2-3	2-3	3.9	3.5
DMAPP	1-2	1-2	n.u.	n.u.
GPP	4-6	4-6	20	24
FPP	n.u.	n.u.	0.9	0.5
Product	FPP	FPP	GGPP	GGPP
Metal iron stimulation	$Mg^{2+}>Mn^{2+}$	$Mg^{2+}>Mn^{2+}$	$Mg^{2+}>Mn^{2+}$	$Mg^{2+}>Mn^{2+}$
pH optimum	6.8	6.8	7.5,9.5	8.5
MW	72,500	72,500	72,000	72,000

[a]Seedlings infected with R. stolonifer.

[b]Abbreviations: IPP = fqayhtpht|1-PP; DMAPP = dimethylallyl-PP; GPP = geranyl-PP; FPP = farnesyl-PP; UUYY % vnoerkdvnoerkd9YYV r.u. = not utilized.

Table 4. Enzyme activities in extracts of castor bean seedlings.

Enzyme	Units [a]/100 seedlings (a) Untreated[b]	(b) 16-hr exposure to Rs spores[c]	(b)-(a)
Geranyl transferase	12	22	10
Farnesyl transferase	0.25	15	~15
Casbene synthetase	0	250	250

[a]nmol product formed/min

[b]Seedlings maintained 67 hr under sterile conditions before extraction and assay.

[c]Seedlings maintained 51 hr under sterile conditions before innoculation with R. stolonifer spore suspension in potato-dextrose medium. Extracts for assay prepared after an additional 16 hr.

occurred prior to measuring the levels of the prenyl transferases. This makes it difficult to compare in a meaningful way the absolute levels of the different enzymes. The higher levels of farnesyl transferase and casbene synthetase present in elicited seedlings can be rationalized in terms of the requirements for casbene synthesis under these conditions. Initially, it was concluded that the higher level of geranyl transferase in elicited seedlings was also related to casbene production, but the results of enzyme compartmentation studies in the following section suggest that this is not the case.

Subcellular localization of enzymes involved in casbene biosynthesis. The subcellular localization of these enzymes in the endosperm of germinating castor bean seed has been investigated by means of the sucrose density gradient procedures of Cooper and Beevers[46]. The cotyledons were removed from a group of seedlings that had germinated for 67 hr under the desired conditions and the pooled endosperms were diced into small pieces with a sharp blade in the presence of buffer containing 20% (w/w) sucrose. The suspension was centrifuged for a short time at 270 *g* and the resulting supernatant suspension was transferred to the top of a linear sucrose gradient ranging from 20% sucrose at the top to 60% sucrose at the bottom. The 270 *g* pellet, which was enriched in proplastids, was resuspended in 20% sucrose-buffer solution and transferred to the top of a 5-step sucrose gradient ranging from 33% sucrose at the top to 60% sucrose at the bottom. Both gradients were centrifuged at 20,000 rpm for 3 hr in a swinging bucket SW-27 rotor. Fractions were collected and analyzed for marker enzymes to locate the positions of the glyoxysomes (catalase), proplastids (triose phosphate isomerase), mitochondria (fumarase), and microsomes (NADH-cytochrome *c* reductase). The fractions were also analyzed for the enzymes of interest and their distributions were correlated with those of the marker enzymes to establish their subcellular localizations.

This analysis was performed with endosperm from pools of seedlings (a) maintained in a sterile environment for the entire 67-hr germination period (sterile), and (b) inoculated with *R. stolonifer* spores 20 hr prior to the termination of the germination period (inoculated). Table 5 presents a qualitative summary of the findings. It should be mentioned that low levels of activity in an organelle could

Table 5. Localization of geranyl transferase and casbene synthetase in sub-cellular organelles of germinating castor bean endosperm.

Enzyme	Glyoxy-somes	Proplas-tids	Mitochon-dria	Micro-somes	Super-natant
Geranyl transferase					
sterile[a]	-	+	-	-	+
inoculated[b]	-	+	+	-	+
Farnesyl transferase					
sterile[a]	-	-	-	-	-
inoculated[b]	-	+	-	-	+
Casbene synthetase					
sterile[a]	-	tr	-	-	tr
inoculated[b]	-	+	-	-	+

+ = present in readily detectable levels
tr = present in very low levels
- = not detected

[a]Seedlings maintained under sterine conditions as noted in Table 4.

[b]Seedlings inoculated with Rs spores as noted in Table 4.

easily be missed, since the samples applied to the gradients were not large. The only one of the three enzymes detected in sterile seedlings at significant levels was geranyl transferase, which was found in the proplastids. This was somewhat unexpected since it has been shown that developing chloroplasts make little or no squalene or sterols from farnesyl-PP[14]. Smaller amounts of farnesyl-PP may be required for synthesis of diterpenes such as kaurene, but this demand must not be high in sterile seedlings since the farnesyl transferase activity was below detectable levels in the proplastid. It may be that proplastids, unlike developing chloroplasts, do synthesize sterols, or perhaps some other unknown product, from farnesyl-PP. After inoculation, all three enzyme activities were readily detected in the proplastid fraction. Readily detectable geranyl transferase activity was also present in the mitochondrion after inoculation; none had been seen in this organelle from sterile seedlings. In all cases in which a substantial amount of an activity was found in association with an organelle, it was also present in the supernatant fractions recovered from the top of the gradient. This is no doubt a consequence, at least in part, of organelle disruption and release of soluble enzymes during the preparation steps. But it is not possible to determine from this type of experiment whether some enzyme was also initially present in the cytosol fraction.

From these results (Table 5) it seems likely that the biosynthesis of casbene in inoculated seedlings occurs in the proplastids. Only the proplastid fraction shows the large increases in farnesyl transferase and casbene synthetase activities that parallel the large increases seen in these enzymes in total extracts. The geranyl transferase activity in the proplastids, on the other hand, did not differ significantly in the sterile and inoculated seedlings. The farnesyl transferase and casbene synthetase activities increase after infection to provide a competing pathway for the proplastid pools of farnesyl-PP produced by the action of geranyl transferase.

The greatly increased levels of geranyl transferase in mitochondria from inoculated seedlings in comparison with sterile seedlings was not expected and its significance is not clear. Since neither farnesyl transferase nor casbene synthetase was detected in mitochondria, this pool of geranyl transferase presumably has nothing to do with diterpene biosynthesis. Perhaps the farnesyl-PP produced there serves as a precursor of sesquiterpenes or triterpenes that have some function in disease resistance. As mentioned earlier, a number of sesquiterpenoid phytoalexins are known in other systems. It was also interesting to find that only one geranyl transferase isoenzyme was detected by ion exchange chromatography of extracts of sterile seedlings, whereas, as pointed out earlier, two isoenzymes were found in extracts of seedlings inoculated with fungal spores. It is possible that one isoenzyme is present in the proplastid under all conditions, whereas the second isoenzyme is specifically elicited in the mitochondrion after exposure to fungi.

These results reinforce the already substantial evidence that compartmentation is an important regulatory factor in terpenoid biosynthesis in plants. It is not surprising to learn that the pathway for biosynthesis of the diterpene casbene is localized in the proplastid. Kaurene synthetase has been shown to be present in proplastids from the endosperm of developing castor bean seed[17] along with isopentenyl-PP isomerase and geranylgeranyl-PP synthetase[47]. Moore and Coolbaugh[16] have shown the presence of kaurene synthetase in extracts of sonicated chloroplasts and Gomez-Navarette and Moore[48] have further provided evidence in favor of the proposal that the biosynthesis of proteins leads to an increased kaurene synthetic capacity in the developing chloroplast during de-etiolation. The developing chloroplast is also the site of synthesis of other types of products such as carotenoids, chlorophylls, and other mixed terpenoids dependent on geranylgeranyl-PP as a substrate[14]. Of course, it should be kept in mind that the proplastids are physiologically distinct from chloroplasts. But it would appear that plastids in general are the organelles involved in the production of geranylgeranyl-PP and its utilization for biosynthesis.

Mechanisms in the regulation of casbene biosynthesis. The studies presented above give some information about the two extremes of the process of elicitation of casbene biosynthesis--inhibition through the action of fungal elicitors and the eventual response through changes in the activities of the participating enzymes. The intervening mechanisms that are operating to translate the triggering events into increased capacity for casbene biosynthesis are not understood at all. Phytoalexin accumulation generally follows a lag time of several hours after elicitation and additional hours before the levels reach a maximum. This type of obser-vation plus other evidence has led to the proposal that *de novo* protein biosynthesis is involved in the expression of the stimulus (see reviews by Keen and Bruegger[44] and Cruickshank[29] for a discussion of some of the evidence for this view). Yoshikawa *et al.*[49] have recently presented some convincing evidence with inhibitors for a requirement for *de novo* messenger-RNA and protein biosynthesis in glyceollin-mediated disease resistance in soybean hypocotyls. But even if *de novo* protein biosynthesis is necessary, there remain many possible ways in which this requirement could be involved. It is probably fair to say that the molecular mechanisms which regulate phytoalexin biosynthesis are still poorly understood; certainly that is the case for casbene biosynthesis.

CONCLUDING REMARKS

We have attempted to illustrate with these examples from our own work on diterpene biosynthesis the likelihood that the types of regulatory mechanisms which operate more generally in biosynthetic metabolism are also important in governing the rates of production of terpenoid compounds in higher plants. Among the general types of regulation seen are the following:

(a) The modulation of enzyme activity by changing concentrations of metabolic effectors. This is illustrated by adenylate energy charge control of the rate of the general pathway of terpenoid biosynthesis. The related type of control by end-product feedback inhibition has not been observed for the examples discussed here. But it has been proposed by McFarlane *et al.*[50] as a controlling feature for the enzyme geraniol hydroxylase, which catalyzes the initial step in the utilization of geraniol or nerol for the biosynthesis of several families of indole alkaloids. One of the end-product alkaloids, cantharanthine, inhibits this enzyme in a reversible manner.

(b) The roles of synthetic substances as artificial regulators of enzyme activities. These are well demonstrated by the potent actions of plant growth retardants such as

Amo-1618 as inhibitors of the A-activity of kaurene synthetase and ancymidol as an inhibitor of kaurene oxidase. Such inhibitions are of interest because of the possibility that they might be mimicking the actions of natural regulatory substances, but in these particular cases no physiological equivalents to the synthetic inhibitors have been discovered so far.

(c) Regulation by virtue of enzyme organization. This is seen at two levels in the work described. There is evidence of increased catalytic efficiency through channeling with the multi-enzyme complex kaurene synthetase. Also the importance of enzyme compartmentation in sub-cellular organelles is emphasized by the demonstration of the localization of casbene biosynthesis in the proplastids in castor bean seedlings.

(d) The relatively slow adjustment in levels of active enzyme as a controlling feature. Although the mechanistic basis for it is not understood, this phenomenon is well illustrated by elicitor control of geranyl transferase, farnesyl transferase, and casbene synthetase levels in castor bean seedlings.

Regulatory mechanisms are characteristically diverse and complex. Certainly our knowledge of the regulation of diterpene biosynthesis in higher plant systems is still very fragmentary. We are convinced that it is an important aspect, not only from the point of view of basic understanding but also if we are to take greater advantage of a plant's potential for the production of a diverse array of interesting and useful substances.

One instance of a situation where knowledge of the natural regulation of isoprenoid metabolism could conceivably be of benefit is with elicitor control of phytoalexin production. It has been suggested that either biotic or abiotic elicitors might be applied exogenously to activate natural defense mechanisms in plants including those that produce terpenoid phytoalexins. (See reviews by Keen and Bruegger[44] and Cruickshank[29] for discussions of some proposed methods for manipulating the phytoalexin response to advantage.) There are still many uncertainties about this approach and much more basic information is needed, but it seems worthy of exploration.

The use of plant hydrocarbons as a petroleum substitute represents another area where the ability to increase the natural production of isoprenoid substances could be advantageous. Calvin, in particular, has argued persuasively for this approach[51]. He has called attention to latex-producing plants from several families that will grow in semiarid areas receiving high levels of solar radiation. In particular, he has chosen *Euphorbia lathyris*, a plant from the same family as the castor bean (Euphorbiaceae), for preliminary investigations. Eight to ten percent of the dry weight of this plant is organic-extractable, hydrocarbon-like material of which one-half is made up of isoprenoid substances including a rubber-like polyisoprenoid and cyclic terpenes, diterpenes, and sterols. An ability to manipulate such plants to increase their yield of isoprenoid substances would clearly be of advantage in this potential application.

ACKNOWLEDGEMENT

The authors wish to thank the American Society of Plant Physiologists for permission to use adaptations of published Figures from *Plant Physiology* for Figures 3, 4, 9 and 11 in this article. Citations to the original articles in which the Figures appeared are included in the text.

REFERENCES

1. Hedden, P., J. MacMillan and B. O. Phinney. 1978. The metabolism of gibberellins. Annu. Rev. Plant Physiol. 29:149-192.
2. Jones, R. L. 1973. Gibberellins: their physiological role. Annu. Rev. Plant Physiol. 24:571-598.
3. Atkinson, D.E. 1969. The enzymes as control elements in metabolic regulation. In "The Enzymes" (P. D. Boyer, ed.) 3rd Ed. 1:461-489. Academic Press, New York.
4. Knotz, J., R. C. Coolbaugh and C. A. West. 1977. Regulation of the biosynthesis of ent-kaurene from mevalonate in the endosperm of immature Marah macrocarpus seeds by adenylate energy charge. Plant Physiol. 60:81-85.
5. Gray, J. C., and R. G. O. Kekwick. 1973. Mevalonate kinase in green leaves and etiolated cotyledons of the French bean Phaseolus vulgaris. Bio-chem. J. 130:983-995.
6. Hanson, J. F. and A. J. White. 1969. Studies in terpenoid biosynthesis. Part IV. Biosynthesis of the kaurenolides and gibberellic acid. J. Chem. Soc. (C):981-985.
7. Shechter, I. and C. A. West. 1969. Biosynthesis of gibberellins. IV. Biosynthesis of cyclic diterpenes from trans-geranylgeranyl pyrophosphate. J. Biol. Chem. 244:3200-3209.
8. Fall, R. R. and C. A. West. 1971. Purification and properties of kaurene synthetase from Fusarium moniliforme. J. Biol. Chem. 246:6913-6928.
9. Frost, R. G. and C. A. West. 1977. Properties of kaurene synthetase from Marah macrocarpus. Plant Physiol. 59:22-29.
10. Buggy, M. J., G. Britton and T. W. Goodwin. 1974. Terpenoid biosynthesis by chloroplasts isolated in organic solvents. Phytochemistry 13:125-129.
11. Shah, D. V., D. H. Feldbuegge, A. R. Houser and J. W. Porter. 1968. Conversion of ^{14}C-labeled geranylgeranyl pyrophosphate to phytoene by a soluble tomato plastid system. Arch. Biochem. Biophys. 127:124-131.
12. Rudinger, W., J. Benz, V. Lempert, S. Schock and D. Steffens. 1976. Inhibition of phytol accumulation with herbicides. Geranylgeraniol and dihydrogeranyl-

geraniol-containing chlorophyll from wheat seedlings. Z. Pflanzenphysiol. 80:131-143.
13. Rudinger, W., P. Hedden, H.-P. Kost and D. J. Chapman. 1977. Esterification of chlorophyllide by geranylgeranyl pyrophosphate in a cell-free system from maize shoots. Biochem. Biophys. Res. Commun. 74:1268-1272.
14. Goodwin, T. W. 1977. In "Lipids and Lipid Polymers in Higher Plants" (M. Tevini and H. K. Lichtenthaler, eds.) pp. 29-47. Springer-Verlag, Berlin.
15. Robinson, D. R. and C. A. West. 1970. Biosynthesis of cyclic diterpenes in extracts from seedlings of Ricinus communis L. II. Conversion of geranylgeranyl pyrophosphate into diterpene hydrocarbons and partial purification of the cyclization enzymes. Biochemistry 9:80-89.
16. Moore, T. C. and R. C. Coolbaugh. 1976. Conversion of geranylgeranyl pyrophosphate to ent-kaurene in enzyme extracts of sonicated chloroplast. Phytochemistry 15:1241-1247.
17. Simcox, P. D., D. T. Dennis and C. A. West. 1975. Kaurene synthetase from plastids of developing plant tissues. Biochem. Biophys. Res. Commun. 66: 166-172.
18. Maudinas, B., M. L. Bucholtz, C. Papastephanou, S. S. Katiyar, A. V. Briedis and J. W. Porter. 1977. The partial purification and properties of a phytoene-synthesizing enzyme system. Arch. Biochem. Biophys. 180:354-362.
19. Dudley, M. W., T. R. Green and C. A. West, unpublished results.
20. Dueber, M. T., W. Adolf and C. A. West. 1978. Biosynthesis of the diterpene phytoalexin casbene: partial purification and characterization of casbene synthetase from Ricinus communis. Plant Physiol. 62:(in press).
21. Simcox, P. D. 1976. The synthesis of kaurene and related diterpenes in higher plants. Ph.D. Dissertation, University of California, Los Angeles.
22. Yafin, Y. and I. Shechter. 1975. Comparison between biosynthesis of ent-kaurene in germinating tomato seeds and cell suspension cultures of tomato and tobacco. Plant Physiol. 56:671-675.
23. Dennis, D. T., C. D. Upper and C. A. West. 1965. An enzymatic site of inhibition of gibberellin biosyn-

thesis by AMO-1618 and other plant growth retardants. Plant Physiol. 40:948-952.

24. Coolbaugh, R. C. and R. Hamilton. 1976. Inhibition of ent-kaurene oxidation and growth by α-cyclopropyl-α-(p-methoxyphenyl)-5-pyrimidine methyl alcohol. Plant Physiol. 57:245-248.

25. Tschabold, E. E., H. M. Taylor, J. D. Davenport, R. E. Hackler, E. V. Krumkalms and W. C. Meredith. 1970. A new plant growth regulator. Plant Physiol. 46:5-19.

26. Coolbaugh, R. C., S. S. Hirano and C.A. West. 1978. Studies on the specificity and site of action of α-cyclopropyl-α-(p-methoxyphenyl)-5-pyrimidine methyl alcohol (ancymidol), a plant growth regulator. Plant Physiol. 62:(in press).

27. Dennis, D. T. and C. A. West. 1967. Biosynthesis of gibberellins. III. The conversion of (-)-kaurene to (-)-kaurene to (-)-kauren-19-oic acid in endosperm of Echinocystis macrocarpa Greene. J. Biol. Chem. 242:3293-3300.

28. Lew, F. T. and C. A. West. 1971. (-)-Kauren-7β-ol-19-oic acid, an intermediate in gibberellin biosynthesis. Phytochemistry 10:2065-2076.

29. Cruikshank, I. A. M. 1977. A review of the role of phytoalexins in disease resistance mechanisms. In "Natural Products and the Protection of Plants." pp. 509-569. Pontificiae Academiae Scientarum Scripta Varia.

30. Ingham, J. L. 1972. Phytoalexins and other natural products as factors in plant desease resistance. Bot. Rev. 38:343-424.

31. Ingham, J. L. 1973. Disease resistance in higher plants. The concept of pre-infectional and post-infectional resistance. Phytopath. Z. 78:314-335.

32. Kuc, J. 1972. Phytoalexins. Annu. Rev. Phytopathol. 10:207-232.

33. Stoessl, A., J. B. Stothers and E. W. B. Ward. 1976. Sesquiterpenoid stress compounds of the Solanaceae. Phytochemistry 15:855-872.

34. Muller, K. O. and H. Borger. 1940. Experimentelle Untersuchungen uber die Phytophthora-Resistant der Kartoffel. Arb. Biol. Reichanst. Landw. Forstwirtsch., Berlin-Dahlem 23:189-231.

35. Muller, K. O. 1958. Studies on phytoalexins. I. The formation and the immunological significance of phy-

toalexin produced by Phaseolus vulgaris in response to infections with Sclerotinia fructicola and Phytophthora infestans. Australian J. Biol. Sci. 11:275-300.

36. Cruikshank, I. 1963. Phytoalexins. Annu. Rev. Phytopathol. 1:351-374.
37. Sitton, D. and C. A. West. 1975. An anti-fungal diterpene produced in cell-free extracts of Ricinus communis seedlings. Phytochemistry 14:1921-1925.
38. Robinson, D. R. and C. A. West. 1970. Biosynthesis of cyclic diterpenes in extracts from seedlings of Ricinus communis L. I. Identification of diterpene hydrocarbons formed from mevalonate. Biochemistry 9:70-79.
39. Crombie, L., G. Kneen and G. Pattenden. 1976. Synthesis of casbene. J. Chem. Soc., Chem. Commun.: 66-68.
40. Cartwright, D., P. Langcake, R. J. Pryce and D. P. Leworthy. 1977. Chemical activation of host defence mechanisms as a basis for crop protection. Nature (London) 267:511-513.
41. Keen, N. T., J. E. Partridge and A. I. Zaki. 1972. Pathogen-produced elicitors of a chemical defense mechanism in soybeans monogenically resistant to Phytophthora megasperma var. sojae. Phytopathology 62:768.
42. Stekoll, M. and C. A. West. 1978. Purification and properties of an elicitor of castor bean phytoalexin from culture filtrates of the fungus Rhizopus stolonifer. Plant Physiol. 61:38-45.
43. Albersheim, P. and A. J. Anderson-Prouty. 1975. Carbohydrates, proteins, cell surfaces, and the biochemistry of pathogenesis. Ann. Rev. Plant Physiol. 26:31-52.
44. Keen, N. T. and B. Bruegger. 1977. Phytoalexins and chemicals that elicit their production in plants. In "Host Plant Resistance to Pests, ACS Symposium Series, No. 62" (P. A. Hedin, ed.). pp. 1-26. Am. Chem. Soc., Washington.
45. Green, T. R. and C. A. West. 1974. Purification and characterization of two forms of geranyl transferase from Ricinus communis. Biochemistry 13:4720-4729.
46. Cooper, T. G. and H. Beevers. 1969. Mitochondria and glyoxysomes from castor bean endosperm. En-zyme constituents and catalytic capacity. J. Biol. Chem. 244:3507-3513.

47. Green, T. R., D. T. Dennis and C. A. West. 1975. Compartmentation of isopentenyl pyrophosphate isomerase and prenyl transferase in developing castor bean endosperm. Biochem. Biophys. Res. Commun. 64:976-982.
48. Gomez-Navarrete, G. and T. C. Moore. 1978. Effect of protein synthesis inhibitors on ent-kaurene biosynthesis during photomorphogenesis of etiolated pea seedlings. Plant Physiol. 61:889-892.
49. Yoshikawa, M., K. Yamauchi and H. Masago. 1978. De novo messenger RNA and protein synthesis are required for phytoalexin-mediated disease resistance in soybean hypocotyls. Plant Physiol. 61:314-317.
50. McFarlene, J., K. M. Madyastha and C. J. Coscia. 1975. Regulation of secondary metabolism in higher plants. Effect of alkaloids on a cytochrome P-450 dependent monooxygenase. Biochem. Biophys. Res. Commun. 66:1263-1269.

CHAPTER SEVEN

COMPOUNDS FROM MICROALGAE--THEIR INFLUENCE ON THE FIELD OF MARINE NATURAL PRODUCTS

YUZURU SHIMIZU

Department of Pharmacognosy
College of Pharmacy
University of Rhode Island
Kingston, Rhode Island

INTRODUCTION

The exploitation of the oceans for new mineral and other resources has only relatively recently become the subject of serious endeavor. The trend has been extended to the pharmaceutical field and the recent surge of interest in natural products from marine organisms has been truly remarkable. Organic chemists have been especially intrigued since many of the compounds isolated have been found to possess new exotic structural features. However, despite the great number of compounds obtained from both marine plants and animals, the fundamental question regarding the true origin of the isolated substances has expectedly been ignored by the majority of chemists.

In the case of terrestrial species, there appears to be a fairly direct correlation between the production of secondary metabolites and the organisms in which the compounds are found. Of course, there are some outstanding exceptions such as the insect sterols which are derived from those present in their food sources. On the contrary, in the aquatic environment, the dependency of one organism on another for the provision of secondary metabolites or their precursors seems to be much more ubiquitous. The transfer of such metabolites takes place essentially in two forms: either _via_ food chains or through symbiosis.

The first clear example of the transfer of secondary metabolites from one organism to another was observed with the sesquiterpenoids isolated from sea hares (_Aplysia_; Gastropoda). In 1965 Hirata's group isolated and determined the structures of aplysin and its analogs from these animals[1]. Later, Irie's group at Hokkaido isolated a series of cognate compounds from red algae, _Laurencia_ sp.[2], and since it is known that the sea hares feed on these algae it seems obvious that aplysin originated from the latter.

Several structurally interesting steroids have been isolated from sponges (Porifera)[3] but it has been suggested that sponges lack the ability to synthesize steroids _de novo_ and instead modify those derived from symbiotic algae. A similar speculation is also made for marine sterols with a 24-norcholestane skeleton, which are found widely distributed in the marine environment[4].

One of the most significant findings in marine natural product chemistry is the discovery of prostaglandin derivatives in the soft coral, _Plexaura homomalla_ (Coelenterata)[5]. An attempt to show that prostaglandin production actually occurs in the symbiotic green algae was negative,[6] but it is still highly probable that some symbiotic or associated organism could play an important role in the production of the compounds. At least there may be a combined synthesis between the host and the symbiont.

It is probably not an overstatement to say that the majority of marine secondary metabolites result directly from, or are influenced by, the interaction of two or more organisms. Micro-unicellular algae, which are responsible for a large portion of production in the ocean, predictably play an important role. Two cases in which we have obtained evidence of such involvement of microalgae in secondary metabolite production are discussed in the rest of this chapter.

STEROLS IN DINOFLAGELLATES

Many marine animals lack the ability to synthesize sterols; nevertheless most of them contain sufficient amounts of these compounds. Gorgonians (soft corals), for example, contain a peculiar sterol, gorgosterol 1 (Figure 1)[7]. Its structure, which has a uniquely substituted side chain with a cyclopropane ring, was established jointly by three groups in 1970[8-10]. Another cyclopropane-ring sterol, acanthasterol, 2, was later isolated from the crown-of-thorns starfish, *Acanthaster planci*[11,12]. The difference between 1 and 2 is merely the position of the double bond in the B ring. It is known that starfish are very effective "chemical machines" that convert the Δ^5-double bond to Δ^7. According, it was assumed that acanthasterol was of dietary origin. Indeed, the starfish are known to feed on corals. Next, was the question of the origin of gorgosterol (1) in gorgonians. The involvement of zooxanthellae, dinoflagellates symbiotic with corals, in gorgosterol production was suspected[13]. A few years ago we had a chance to examine the sterol content of the dinoflagellate, *Gonyaulax tamarensis*, which had been mass-cultured in connection with a study of its toxicity. The major sterol in this organism was found to be the C_{30} sterol which we named dinosterol 3. The structural study revealed that the sterol is 4α,23,24ξ-trimethyl-5α-cholest-22-en-3β-ol[14]. The uncommon 23-methyl side-chain of 3 strongly implied its close relationship with gorgosterol (1), and thus suggested that dinoflagellates were a possible source of gorgosterol in the gorgonians. An X-ray crystallographic study of dinosterol *p*-iodobenzoate assigned the *E* form to the Δ^{22} double bond and identical C-24 absolute configuration (*R*) in both dinosterol and gorgosterol[15]. The result provided evidence that dinosterol was either a possible direct precursor or a compound just off the main pathway leading to gorgosterol (Scheme 1). The next question was where the formation of the cyclopropane ring takes place. I believe that it occurs in the dinoflagellate and we have been able to detect a small amount of a C_{31} sterol in the *G. tamarensis* extract[16]. Thus, in this particular case, we were able to demonstrate the probable proliferation of secondary algal metabolites in three organisms from their structural likeness.

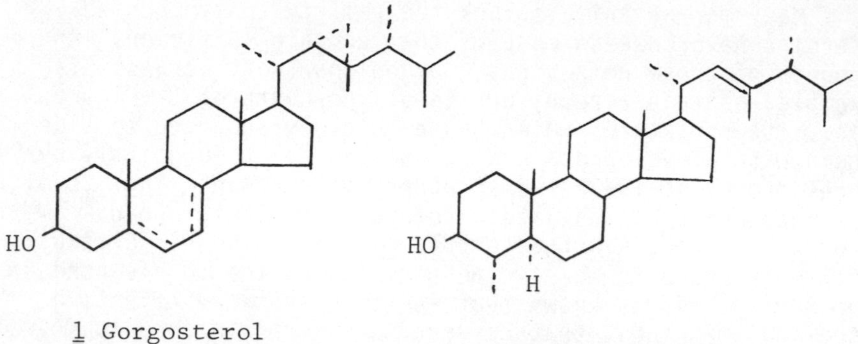

Figure 1. Structures of gorgosterol from gorgonians and dinosterol from dinoflagellates.

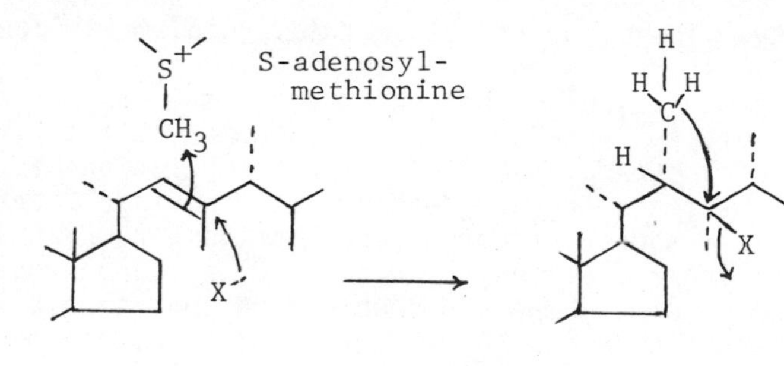

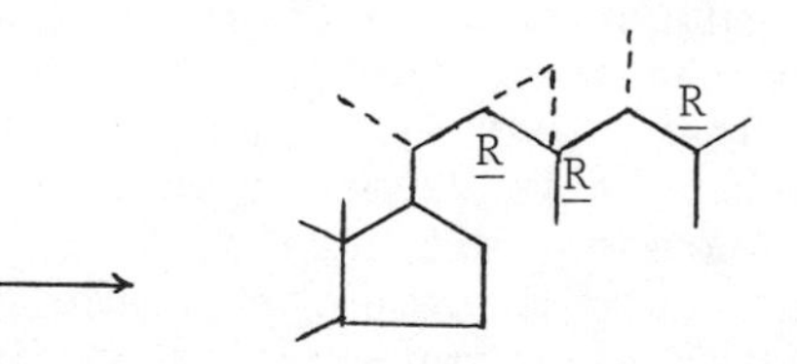

Scheme 1. Possible mechanism for the formation of gorgosterol side chain from dinosterol type side chain.

MARINE TOXINS AND DINOFLAGELLATES

Marine organisms are very rich sources of various types of compounds which are toxic to a variety of other living forms. Unique types of potent toxins such as tetrodotoxin, saxitoxin, and palytoxin are derived from marine organisms. The occurrence of such an abundance and diversity of toxins in marine animals may not be coincidental in view of the highly competitive environment in which these organisms exist. However, very little is known about the biogenetic origins of the toxins which they produce.

An elaborate example of toxin transfer in the marine environment has been recently demonstrated. In 1976 Kato and Scheuer at the University of Hawaii isolated and successfully elucidated the structure of the toxin aplysiatoxin, 4, found in the sea hare, *Stylocheilus longicauda*[17,18]. Later, Moore *et al*. at the same institution isolated the debromo derivative 5 of this very characteristic molecule from *Lyngbya* spp. (Cyanophyta)[10]. It was observed that the sea hare was feeding on the blue-green alga, and there is thus little question about the true origin of the basic skeleton of aplysiatoxin.

Probably the most spectacular example of metabolite transfer in the food chain is in paralytic shellfish poisoning (PSP) (Figure 3), which is of great interest because of the high toxicity to man of the compounds involved. Until recently only one toxin, saxitoxin (6) (Figure 4), was identified and a simple straightforward scheme was accepted for the relationship between the causative organism and toxic shellfish: the saxitoxin-producing dinoflagellates, *Gonyaulax* spp. (red tide), are ingested by filter-feeding shellfish and the toxin then accumulates in their flesh[20a]. The major source of saxitoxin was known to be Alaska butter clams[20b].

The author's group started work on this subject in 1972, initially to identify the paralytic toxin in North Atlantic shellfish. We soon realized that the problem was much more complicated than first assumed. First of all, several toxins were found to be involved in PSP, saxitoxin being only one component. Indeed, it is completely absent from some toxic shellfish. Secondly, even when morphologically indistinguishable, the *Gonyaulux* spp. varied from non-

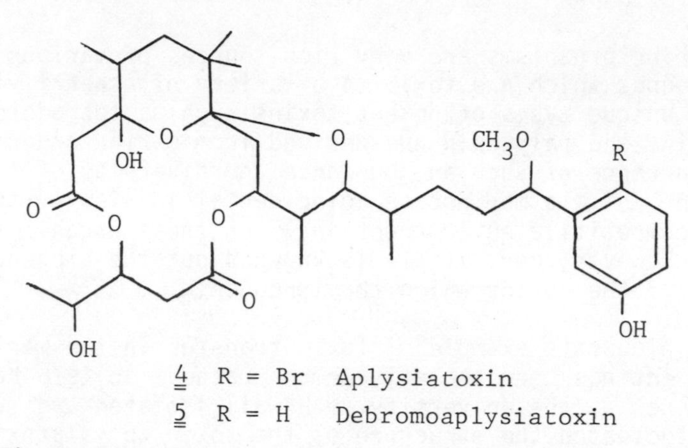

Figure 2. Structures of aplysiatoxin from sea hares and debromoaplysiatoxin from blue-green algae.

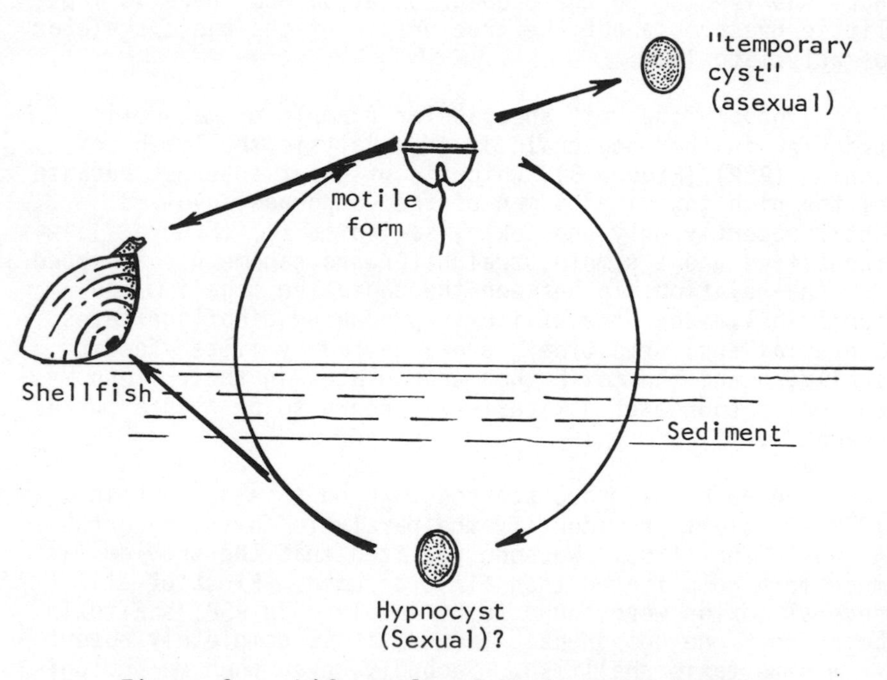

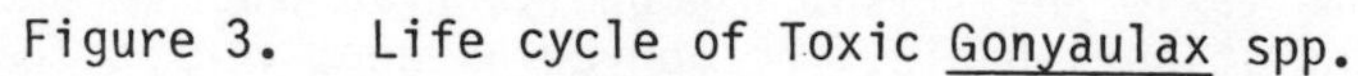

Figure 3. Life cycle of Toxic Gonyaulax spp.

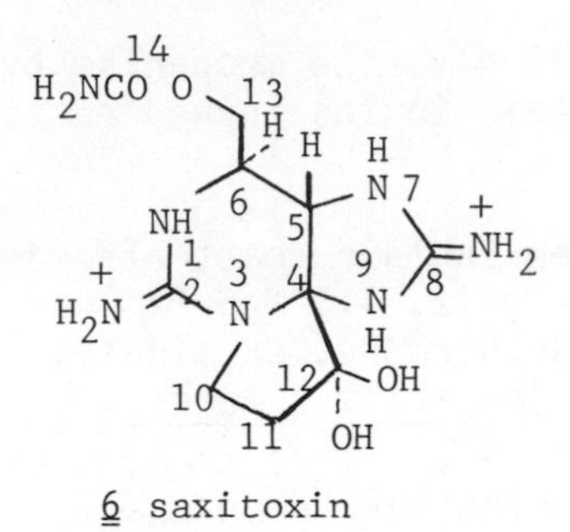

6 saxitoxin

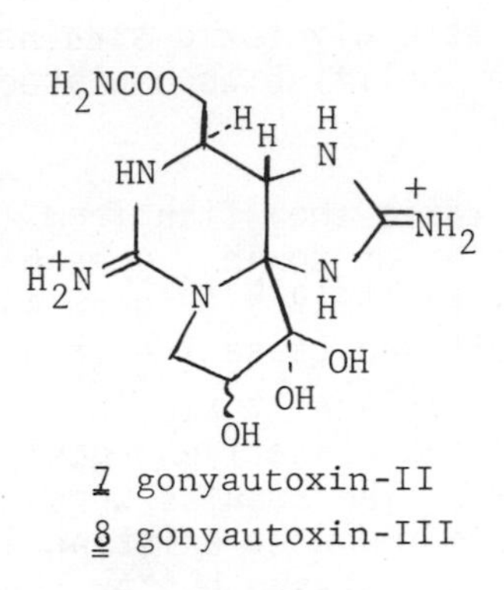

7 gonyautoxin-II
8 gonyautoxin-III

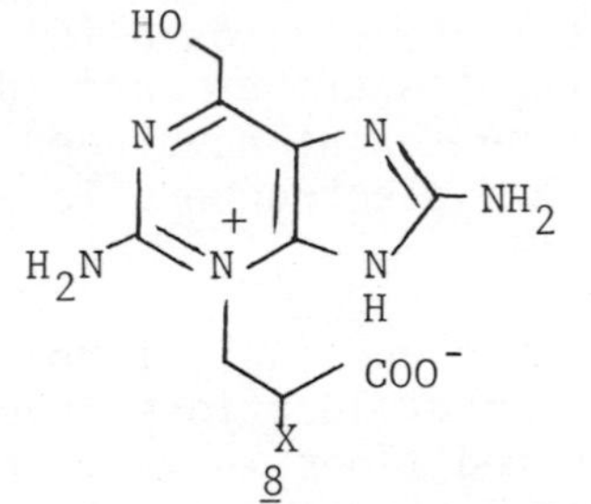

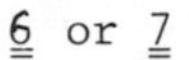

H_2O_2-NaOH →

8
+

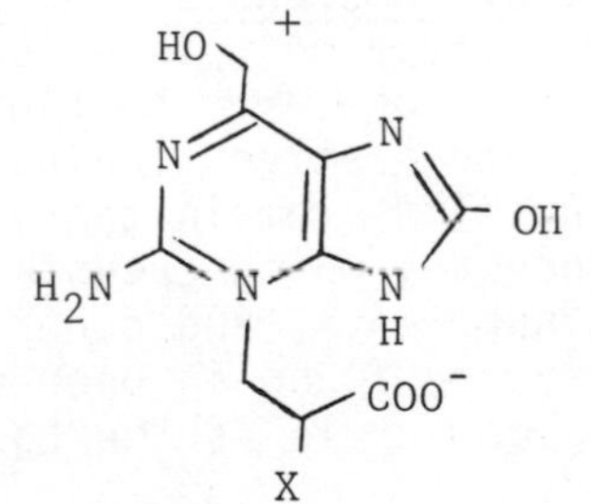

9
X = H from saxitoxin
X = OH from gonyautoxin-II

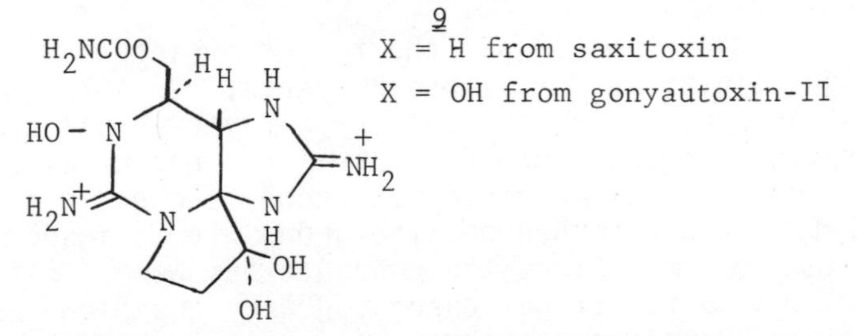

10 neosaxitoxin

Figure 4. Toxins from Gonyaulax spp. and their degradation products.

toxic to strongly toxic strains. Thirdly, the mechanism by which the shellfish accumulated certain toxins turned out to be unclear.

The toxic shellfish from the New England coast afforded a new series of toxins, gonyautoxin-I, II, III and IV, besides saxitoxin[21a,b] (Figure 4). Furthermore, examination of the cultured causative organism, Gonyaulax tamarensis, gave an additional toxin, neosaxitoxin[22]. Since then, many shellfish and causative organism samples have been analyzed[23-27]. The results were always the same, and several toxins were found co-existing in each specimen. The only exception was Alaska butter clam, which was found to contain almost exclusively saxitoxin (6) and a trace of neosaxitoxin (11). The results are summarized in Tables 1 and 2. So far a total of nine toxins have been identified and their chromatographic and electrophoretic properties are summarized in Table 3.

Cultivation and analysis of several strains of Gonyaulax from different locations revealed an intriguing aspect of the toxic red tides[2,28,29]. Apparently, with both G. catenella and G. tamarensis, there are some chemical strains of morphologically almost identical individuals whose toxicity ranges from non-toxic to highly toxic. In confined waters, there are "native strains" which are indigenous to each location. They remain there in sediments as sexual dormant hypnocysts, which excyst and multiply asexually when temperature, nutrients, and other factors become optimum. Sometimes they are found in open waters, but the hypnocyst-motile form-bloom cycle is the basis of regular red tide occurrences[30].

Saxitoxin (6) was the target of intensive structural investigations for almost 20 years. In 1975 the final structure, 6, was obtained by X-ray crystallography of saxitoxin p-bromobenzenesulfonate[31] and hemiketal hydrochloride[32a,b]. The molecule was shown to contain a hydrated ketone and a perhydropurine group with a fused five-membered ring system. Since the gonyautoxins were first isolated by a combination of gel chromatography and high-performance ion-exchange chromatography[21], efforts have been made to establish their structures. However, there are two obstacles which prevent the use of ordinary procedures for structure elucidation: (a) the highly hygroscopic nature of the

Table 1. Toxin Contents in Various Shellfish Samples

Toxic Shellfish	Causative Organism*	Collected Site	Purefied Toxins+								
			STX	GTX_1	GTX_4	GTX_2	GTX_3	GTX_5	neoSTX	GTX_6	GTX_7
Mya arenaria	*G. tamarensis*	Essex, MA	++	++	+	+++	+		+		
Mya arenaria	*G. tamarensis*	Hampton, MA	++	±		+++	+		+		
Mytilus edulis	*G. catenella*	Oase Bay, Mie, Japan	+	+++	+	++	+	++			
Tapes japonica	*G. catenella*	Oase Bay, Mie, Japan	+	+++	+	++	+	++		+	
Mytilus edulis	*Gonyaulax* sp.	Vigo, Spain	+++	+	+	++	+	++		+	
Mytilus edulis	*G. catenella* ?	Haines, AK		+++	++	++	++	+	+	+	
Mytilus edulis	*G. catenella* ?	Elfin Cove, AK	+	+++	+	+	++	+		+	
Saxidomus giganteus	?	Porpoise Island, AK	+++						+		
Placopecten magellanicus	*G. tamarensis*	Bay of Fundy, N.S., Canada	+	++	+	++	++	+	+++		+
Mytilus, spp.	*G. catenella*	Inland Sea, Japan	+			+++	++	++			

*Tentative assignment: all *Gonyaulax* spp.

+See Figure 3: STX = saxitoxin ; GTX, etc. = Gonyautoxins I etc.; neo STX = neosaxitoxin.

Table 2. Toxin Profile in *Gonyaulax* Organisms of Different Origins

Organisms	Purified Toxins							
	STX	GTX_1	GTX_2	GTX_3	GTX_4	GTX_5	neoSTX	Toxicity
G. tamarensis[1]	++	++	+++	+	+	+++	+++	1 mu/10^4cells
G. tamarensis[2]	+	++	++	++	+	+	+	1 mu/5×10^4cells
G. tamarensis[3]	±	++	++	++	+	+	+	1 mu/10^5cells
G. catenella[4]	+	+++	++	+	+	++		1 mu/4×10^4cells
G. catenella[5]	++	++	++	++	+	+	++	1 mu/10^5cells
G. catenella[6]			(±)					1 mu/10^6cells

[1]Ipswich, MA; [2]Mill Pond, Cape Cod, MA; [3]Perch Pond, Cape Cod, MA; [4]Oase Bay, Japan; [5]Sequin Bay, WA; [6]LaJolla, CA.

Table 3. Chromatographic and Electrophoretic Behavior of Isolated Toxins

Toxins	R_f[1]	R_f[2]	Electro-phoresis Rm[3]	Elution Order from BioRex 70 Column
STX	0.62	0.51	1.00	9
GTX_1	0.90	0.70	0.16	4
GTX_4	0.81	0.65	0	3
GTX_2	0.81	0.65	0.56	6
GTX_3	0.69	0.61	0.28	5
GTX_5	0.61	0.52	0.28	2
neoSTX	0.70	0.54	0.47	7
GTX_6	0.57	0.08	1	
GTX_7	0.44		0.97	8
$APTX_1$	0.81	0.51	-	2
$APTX_2$	0.70	0.51	-	3
$APTX_3$	0.48	0.50	1.51	1

[1]Silica gel 60; pyridine, ethyl acetate, water, acetic acid (75:25:30:15)

[2]Silica gel GF; t-butyl alcohol, acetic acid, water (2:1:1)

[3]Relative mobility on cellulose acetate strip in Tris-HCl buffer pH 8.7 at 200 V (constant) and 0.2 mA/cm for 1 hr. The value was calculated by assuming saxitoxin as 1.0.

Table 4. Comparison of NMR Chemical Shifts of Toxins from *Gonyaulax* spp.

C	Saxitoxin		Gonyautoxin-II		Neosaxitoxin	
	CMR	PMR	CMR	PMR	CMR	PMR
2	157.9*(s)	---	158.0*(s)	---	158.8*(s)	---
8	156.1*(s)	---	156.2*(s)	---	158.1*(s)	---
14	159.0*(s)	---	159.1*(s)	---	---	---
4	82.6 (s)	---	81.5 (s)	---	82.2 (s)	---
5	57.3 (d)	4.77 (J=1)	57.9 (d)		56.9 (d)	4.83 (s)
6	53.2 (d)	3.87 (J=9,5,1)	53.2 (d)		64.4 (d)	4.15 (J=6,6)
10	43.0 (t)	3.85 (J=10) 3.57 (J=10)	50.9 (t)		43.7 (t)	3.80 (J=10) 3.58 (J=10)
11	33.1 (t)	2.37 (m)	77.6 (d)		31.9 (?)	2.44 (m)
12	98.9 (s)	---	97.5 (s)	---	98.6 (s)	---
13	63.3 (t)	4.27 (J=11,9) 4.05 (J=11,5)	63.3 (t)		61.1 (t)	4.43 (J=11,6) 4.28 (J=11,6)

*Assignments may be interchanged.

toxins; and (b) the small amounts obtained. For these reasons structural elucidation had to rely on degradations and spectroscopic measurements using micro amounts of samples.

Gonyautoxin-II, 7, and gonyautoxin-III, 8 (Figure 4), were found to be an epimeric pair which equilibrate slowly in solution. Gonyautoxin-II can be aromatized to purinylpropionic acid derivatives, 8 and 9, by oxidation with alkaline H_2O_2 indicating the presence of a perhydropurine skeleton as found in saxitoxin[33]. CMR and PMR data also suggested the structural closeness of gonyautoxin-II and saxitoxin. However, critical differences were observed with the CMR signal of C-11 and PMR signals of hydrogens around this particular location (see Table 4), which strongly implied the presence of a substituent at C-11 of saxitoxin. The presence of an additional substituent was further supported by the fact that the degradation product has a chiral center as evidenced by its CD spectrum. It was speculated that the substituent was a hydroxyl group from the analysis of chemical shifts of gonyautoxin-II and the mass spectrum of the degradation products.

Gonyautoxin-III can now be assigned structure as the 11-epimer of gonyautoxin-II. It was known that saxitoxin undergoes a slow deuterium exchange in plain deuterated water[1]. The same enolization of gonyautoxin-II should result in the epimerization of the 11-hydroxyl function. The equilibrium favors gonyautoxin-II in a ratio of 7:3. Considering steric hindrance, the 11α and 11β form are assigned to gonyautoxin-II and -III respectively.

Another newly found toxin, neosaxitoxin, was first found as a minor toxin in Alaska butter clams. But later it was discovered that "the saxitoxin fraction" from cultured Gonyaulax tamarensis is mostly composed of neosaxitoxin[28]. The chromatographic behavior of saxitoxin and neosaxitoxin is so close that special precautions are necessary to separate the two compounds. To make the matter more complicated, the IR spectra of the compounds are practically identical[34].

Again, micro-spectroscopic and -chemical techniques had to be utilized for the structural study. Unlike saxitoxin, and gonyautoxin-II, neosaxitoxin is not aromatized readily by oxidation. However, the PMR and CMR data suggested that both structures were closely related. There are significant differences in the chemical shifts with C-6, C-5, C-13 and the surrounding hydrogens. Microtitration established that the molecule had three pK_a's. Study of the chemical shift dependency on pH showed that the first pK_a is associated with a charge transfer near C-6.

Reduction of neosaxitoxin with Zn-HCl, Zn-AcOH, or $NaBH_4$ afforded saxitoxin, which was further reduced to dihydrosaxitoxin on prolonged treatment. Among several possibilities, the hydroxyguanidine structure (10) is considered as being likely for neosaxitoxin[35]. Dissociation of the hydroxyl group of the N-oxide form can explain the first pK_a and the changes in chemical shifts associated with it. Neosaxitoxin's difficulty in aromatization also suggests the presence of a substituent on the perhydropurine nucleus. Neosaxitoxin undergoes a very rapid deuterium exchange (< 1 hr) at C-11 in deuterated water. This rate of exchange is much greater than that of saxitoxin implying an enhanced ketonic character of the hydrated carbonyl group in neosaxitoxin. Two strongly electron-withdrawing guanidinium groups stabilize the hydrated form of ketone in saxitoxin. Substitution on one of the guanidinium groups is expected to change the effect. The structures of other toxins (Table 3) are currently being investigated.

MISSING LINKS

As mentioned earlier, saxitoxin is found in the Alaska butter clam, Saxidomus giganteus. The way in which the Alaska butter clams accumulate the toxin is unique; saxitoxin is concentrated in the siphon and stays there year round[36]. Our analysis of toxic mussel samples, however, shows the presence of several toxins which are mainly in the hepatopancreas[26]. Generally, the toxicity of shellfish coincides with an increase in the number of dinoflagellates in their environment and disappears 2-3 weeks after the end of the algae blooms. A completely different mechanism must, therefore, be involved in the case of butter clams. One highly probably way is that the toxic dinoflagellate are symbiotic within the shellfish. This is known with other dinoflagellates and the red pigment peridin, a xanathophyll carotenoid highly specific to dinoflagellates, is found in shellfish meat.

Saxitoxin is also found in a high concentration in the south Pacific crabs, Zosimus aeneus, Platypodia granulosa and Atergatis floridus[37]. Considering their carnivorous nature, the crabs can be considered as forming the secondary consumers in the food chain in which the toxin originates from the dinoflagellates. Of course, symbiosis is another possibility to account for this unexpected occurrence of saxitoxin. There is strong speculation that the mysterious origin of ciguatera toxin in tropical waters is also a dinoflagellate. At the moment, however, there is insufficient evidence available for significant discussion.

The fate of compounds incorporated into animal bodies seems to be very diverse. The presence of only saxitoxin in Alaska butter clams or in the Pacific crabs is clearly a result of biotransformation and/or degradation, because our studies of various dinoflagellate organisms indicate that they always produce several toxins. Already we have discovered that soft-shell clams, Mya arenaria, convert neosaxitoxin into saxitoxin, and thus we see only a trace of neosaxitoxin in these organisms.

CONCLUSION

The examples presented above are literally "drops in the ocean" but may be enough to show the complexity and problems inherent in marine natural product chemistry. More often than not, analogies from the terrestrial organisms are not applicable. The use of chemotaxonomic concepts may sometimes be dangerous as far as marine organisms are concerned. For example, the fact that certain coelenterates afford specific terpenes may not indicate their taxonomical closeness, but rather their similar eating habits. I feel that it is extremely important to understand the chemical entities of marine organisms as a manifestation of their highly interactive living patterns.

ACKNOWLEDGMENTS

The work done at the University of Rhode Island was supported by PHS Grant FD 00619 and Sea Grant, U.R.I., which are greatly appreciated.

REFERENCES

1. Yamamura, S. and Y. Hirata. 1963. Structures of aplysin and aplysinol, naturally occurring bromo compounds. Tetrahedron 19:1485-1496.
2. Irie, T., M. Suzuki, E. Kurosawa, and T. Masumune. 1966. Lauriterol and debromolaurentirol, constituents from Laurencia intermedia. Tet. Lett.:1837-1840.
3. Minale, L. and G. Sodano. 1977. Non-conventional sterols of marine origin. In "Marine Natural Products Chemistry" ed. D. J. Faulkner and W. H. Fenical, Plenum Press, New York, pp. 87-109.
4. Metayer, A. and M. Barbier. 1973. Synthese du 5α-dimethyl-24 choline-22 trans ol-3β. C. R. Acad. Sci. Paris C 276,201-203.
5. Weinheimer, A. J. and R. L. Spraggins. 1969. Two new prostaglandins isolated from the gorgonian Plexaura homomalla (Esper) in "Food-Drugs from the Sea Proceedings" H. W. Youngken, Jr., ed., Marine Technology Society, Washington, D. C., pp. 311-314.
6. Corey, E. J. and W. N. Washburn. 1974. The role of the symbiotic algae of Plexaura homomalla in prostaglandin biosynthesis. J. Am. Chem. Soc. 96:934-935.
7. Bergman, W., M. J. McLean and D. Lester. 1943. Contributions to the study of marine products-- XIII. Sterols from various invertebrates. J. Org. Chem. 8:271-282.
8. Gupta, K. C. and P. J. Scheuer. 1969. Zoanthid sterols. Steroids 13:343-356.
9. Hale, R. L., L. Leclercq, B. Tursch, C. Djerassi, R. A. Cross, A. J. Weinheimer, K. Gupta, and P. J. Scheuer. 1970. Demonstration of a biologically unprecedented side chain in the marine sterol, gorgosterol. J. Am. Chem. Soc. 92:2179-2180.
10. Ling, N. C., R. L. Hale and C. Djerassi. 1970. The structure and absolute configuration of the marine sterol gorgosterol. J. Am. Chem. Soc. 92:5281-5282.
11. Gupta, K. C. and P. J. Scheuer. 1968. Echinoderm sterols. Tetrahedron 24:5831-5837.
12. Sheikh, Y. M., C. Djerassi and B. M. Tursch. 1971. Acaninasterol: A cyclopropane-containing marine sterol from Acanthaster planci. Chem. Commun. 217-218.
13. Ciereszko, L. S., M. A. Johnson, R. W. Schmitz, and C. B. Koons. 1968. Chemistry of coelenterate VI. Occurrence of gorgosterol, a C_{30} sterol in coelen-

terates and their zooxanthellae. Comp. Biochem. Physiol. 24:899-904.

14. Shimizu, Y., M. Alam, and A. Kobayashi. 1976. Dinosterol, the major sterol with a unique side chain in the toxic dinoflagellate, Gonyaulax tamarensis. J. Am. Chem. Soc. 98:1059.
15. Finer, J., J. Clardy, A. Kobayashi, M. Alam and Y. Shiizu. 1977. Identity of the stereochemistry of dinosterol and gorgosterol side chain. J. Org. Chem. 43:1990-1992.
16. Shimizu, Y. and M. Alam, unpublished data.
17. Kato, Y., and P. J. Scheuer. 1975. The aplysiatoxins. Pure Appl. Chem. 41:1-14.
18. Kato, Y., and P. J. Scheuer. 1976. The aplysiatoxins: Reactions with acid and oxidants. Pure Appl. Chem. 48:27-33.
19. Mynderse, J. S., R. E. Moore, M. Kashiwagi and T. R. Norton. 1977. Antileukemia activity in the Oscillatoriaceae: Isolation of debromomaplysiatoxin from Lyngbya. Science 196:538-540.

20a. Sommer, H. and K. F. Meyer. 1937. Paralytic shellfish poisoning. Arch. Pathol. 24:560-598.

20b. Schantz, E. J., J. D. Mold, D. W. Stanger, J. Shavel, F. J. Riel, J. P. Bowden, J. M. Lynch, R. S. Wyler, B. Riegel and H. Sommer. 1957. Paralytic shellfish poison. VI. A procedure for the isolation and purification of the poison from toxic clams and mussel tissues. J. Am. Chem. Soc. 79:5230-5235.

21a. Shimizu, Y., M. Alam, Y. Oshima and W. E. Fallon. 1975. Presence of four toxins in red tide infested clams and cultured Gonyaulax tamarensis cells. Biochem. Biophys. Res. Commun. 66:731-737.

21b. Shimizu, Y., L. J. Buckley, M. Alam, Y. Oshima, W. E. Fallon, H. Kasai, I. Miura, V. P. Gullo, and K. Nakanishi. 1976. Structures of gonyautoxin-II and -III from the east coast toxic dinoflagellate Gonyaulax tamarensis. J. Am. Chem. Soc. 98:5414-5416.

22. Oshima, Y., L. J. Buckley, M. Alam and Y. Shimizu. 1977. Heterogeneity of paralytic shellfish poisons. Three new toxins from cultured Gonyaulax tamarensis cells, Mya arenaria and Saxidomus giganteus. Comp. Biochem. Physiol. 57:31-34.
23. Oshima, Y., W. E. Fallon, Y. Shimizu, T. Noguchi and Y. Hashimoto. 1976. Toxins of the Gonyaulax sp. and infested bivalves in Oase Bay. Nippon Suisan Gakkaishi 42:851-856.

24. Oshima, Y., Y. Shimizu, S. Nishio, and T. Okaichi. 1978. Identification of paralytic shellfish toxins in shellfish from inland sea. Nippon Suisan Gakkaishi 44:395.
25. Hsu, C. P., Y. Shimizu, G. Hunyady, S. Hasler, J. Luthy, U. Zweifel and Ch. Schlatter. 1978. Vergiftungsfaelle durch Miesmuscheln in der Schweiz. 1976. Sanitaetspolizeiliche Massnahmen und Analyse der PSP-Toxine. Mitteilungen aus dem Gebiete der Lebensmitteluntersuchung und Hygiene, in press.
26. Shimizu, Y., W. E. Fallon, J. C. Wekell, D. Gerber, Jr., and E. Gauglitz, Jr. 1978. Analysis of toxic mussels (Mytilus sp.) from the Alaskan inside passage. J. Agric. Food 26 (in press).
27. Hsu, C. P., A. Marchand, and Y. Shimizu. 1978. Paralytic shellfish toxins in the sea scallop, Placopecten magellanicus. in the Bay of Fundy. J. Fish. Res. Board Canada, (in press).
28. Alam, M. I., C. P. Hsu and Y. Shimizu. 1978. Comparison of toxins in three isolates of Gonyaulax tamarensis (dinophyceae) 1979. J. Phycol. 15.
29. Hsu, C. P., M. I. Alam, and Y. Shimizu, unpublished data.
30. Anderson, D. M. and D. Wall. 1978. The potential importance of benthic cysts of Gonyaulax tamarensis and Gonyaulax excavata in initiating toxic dinoflagellate blooms in the Cape Cod region. J. Phycol. 14:224-234.
31. Schantz, E. J., V. E. Ghazarossian, H. K. Schnoes, F. M. Strong, J. P. Springer, J. O. Pezzanite, and J. Clardy. 1975. The structure of saxitoxin. J. Am. Chem. Soc. 97:1238-1239.
32a. Bordner, J., W. E. Thiessen, H. A. Bates and H. Rapoport. 1975. The structure of a crystalline derivative of saxitoxin. The structure of saxitoxin. J. Am. Chem. Soc. 97:6008-6012.
32b. Wong, J. L., R. Oesterlin and H. Rapoport. 1971. The structure of saxitoxin. J. Am. Chem. Soc. 93:7344-7345.
33. Wong, J. L., M. S. Brown, K. Matsumoto, R. Oesterlin and H. Rapoport. 1971. Degradation of saxitoxin to a purimido(2,1-b) purine. J. Am. Chem. Soc. 93:4633-4634.
34. Shimizu, Y., M. Alam, Y. Oshima, L. J. Buckley, W. E. Fallon, H. Kasai, I. Miura, V. P. Gullo and K.

Nakanishi. 1977. Chemistry and distribution of deleterious dinoflagellate toxins, in "Marine Natural Products Chemistry" D. J. Faulkner and W. H. Fenical, ed., Plenum Press, New York, pp. 261-269.

35. Shimizu, Y., C. P. Hsu, W. E. Fallon, Y. Oshima, I. Miura, and K. Nakanishi. 1978. The structure of neosaxitoxin. J. Am. Chem. Soc. (in press).

36. Schantz, E. J. and H. W. Magnusson. 1964. Observations on the origin of the paralytic poison in Alaska butter clam. J. Protozool. 11:239-242.

37. Konosu, S., A. Inoue, T. Noguchi, and Y. Hashimoto. 1968. Comparison of crab toxin with saxitoxin and tetrodotoxin. Toxicon 6:113-117.

Chapter Eight

MOLECULAR ASPECTS OF HALOGEN-BASED BIOSYNTHESIS OF MARINE NATURAL PRODUCTS

WILLIAM FENICAL

Institute of Marine Resources
Scripps Institution of Oceanography
La Jolla, California

INTRODUCTION

An exceptional feature of many marine-derived secondary metabolites is the high degree of halogen-atom substitution, particularly with the element bromine[1]. Halogenated terpenes and acetate-derived compounds are particularly common metabolites of the marine red algae or seaweeds (Rhodophyta), and it is generally considered that these secondary compounds are produced as defensive substances to deter prospective predators[1].

The overall structures and halogen substitution patterns of these compounds, coupled with the results of recent biomimetic synthesis studies[2,3], illustrate that the equivalent of an enzymatically produced bromonium ion (Br^+) reacts initially with typical fatty acid precursors or, in the case of terpenes, with the acyclic precursors geraniol, farnesol, etc. This reaction generally yields halogen-containing metabolites, which are the simple brominated analogs of well known acetate-derived structures or cyclic terpenoid skele-

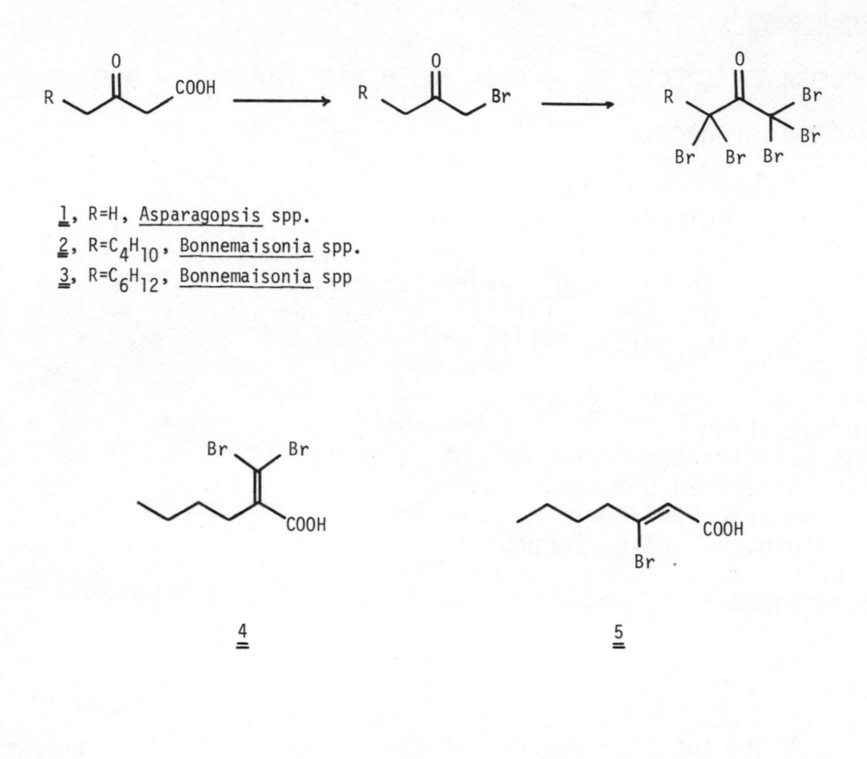
O
COOH
R
O
Br
R
O
Br
Br
Br
Br
Br
R
1, R=H, Asparagopsis spp.
2, R=C4H10, Bonnemaisonia spp.
3, R=C6H12, Bonnemaisonia spp
Br
Br
COOH
COOH
Br
4
5

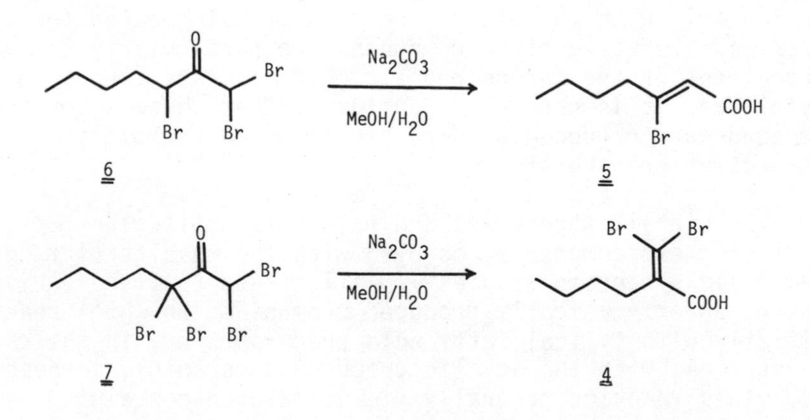
O
Br
Br
Br
Na2CO3
MeOH/H2O
COOH
Br
6
5
O
Br
Br
Br
Br
Na2CO3
MeOH/H2O
Br
Br
COOH
7
4

tons. In this chapter, I point out that these substances are much less stable than their nonbrominated counterparts, and in many cases undergo further transformations involving base reactions and halogen solvolysis. These reactions result in molecular rearrangements to produce new skeletal types which are unprecedented from terrestrial sources.

HALOGENATION AND REARRANGEMENT OF FATTY ACID-DERIVED COMPOUNDS

Red seaweeds of the family Bonnemaisoniaceae are a unique group that produces halogenated secondary metabolites totally via acetate-derived biosynthesis. Based upon the recent halogenating enzyme studies of the Hager group[4], it appears that in Asparagopsis and Bonnemaisonia species homologous acetate-derived precursors such as 1 - 3 are brominated, with concomitant decarboxylation, to yield halomethyl ketones which are themselves prone to further halo-genation. In Asparagopsis species, major amounts of halo-genated acetones are found, while in Bonnemaisonia species brominated 2-heptanones and 2-nonanones are observed[5,6]. In B. hamifera, for example, 1,1,3,3-tetrabromo-2-heptanone is a major metabolite responsible for the fragrant odor of the alga[7]. Careful investigation of the halogen-containing compounds from both B. hamifera and B. nootkana have recently revealed, however, that major amounts of carboxylic acids are produced[6]. 3,3-Dibromo-2-n-butylacrylic acid (4) and 3-bromo-2-heptenoic acid (5) were isolated from B. nootkana, and 5 was observed as the major brominated metabolite in B. hamifera. The C_7 acid, 5, could be a simple metabolic product of acetate biosynthesis, but the branched skeleton (2-methyl-hexane) of compound 4 is quite unusual and difficult to rationalize. Methyl-branched fatty acids do occur in nature, and the "extra" methyl group is generally thought to be derived from the incorporation of an extraneous propionate unit or via alkylation with S-methylmethionine[8]. In this case, McConnell and Fenical[6] have provided convincing evidence that the branched acid 4 is produced by neither of these classical routes, but rather from in vivo base reactions of tri- and tetrabromo-2-heptanones. Treatment of 1,1,3-tribromo-2-heptanone (6) with sodium carbonate in aqueous methanol yielded exclusively the cis bromo acid 5. In consistent fashion, treatment of 1,1,3,3-tetrabromo-2-heptanone (7) under the same conditions gave exclusively the

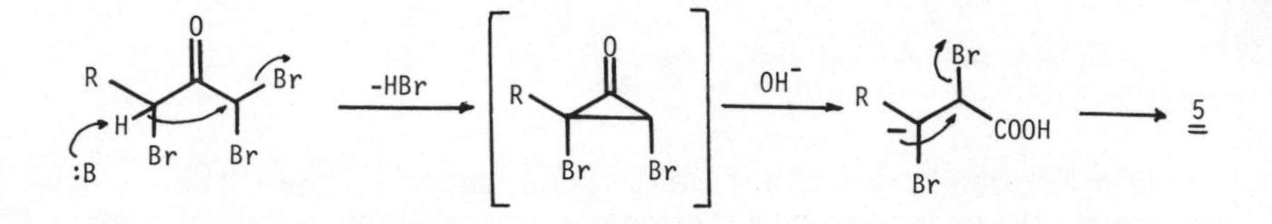

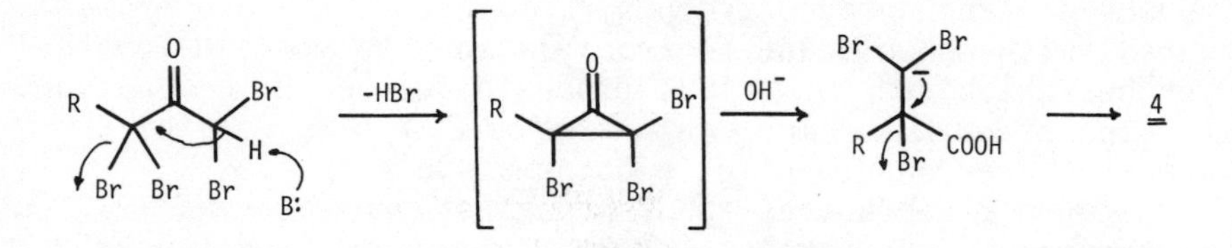

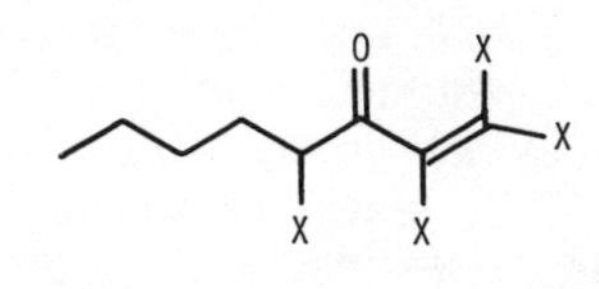

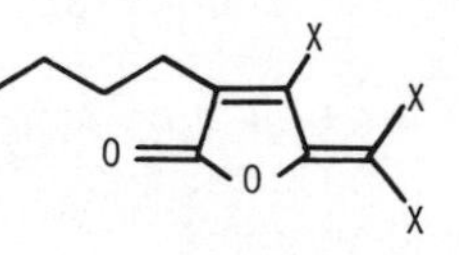

X = Br, Cl, I

8

9

branched dibromo acid 4. Based upon these observations we have proposed[6] that brominated ketones undergo Favorsky rearrangements yielding both linear -enoic acids and branched systems such as compound 4. As outlined below, the selective production of 4 and 5 can be rationalized by ring-opening reactions of a classical cyclopropanone intermediate.

Two further types of compounds, the C_8 halogenated 1-octen-3-one structure, as in 8, and the C_9 halogenated, branched lactone skeleton as in 9 (a 4-methyl-octane system), have been isolated from members of the family Bonnemaisoniaceae (Bonnemaisonia[9,10] and Delisea[11] species). Consideration of the probable mechanism of production of 4 and 5, and that C_9 linear halogenated ketones have been isolated from Bonnemaisonia, prompted the proposal[6] that both octenones and C_9 lactones arise via similar Favorsky reactions from a pentahalo-2,4-nonamedione (10) precursor. As illustrated in Figure 1, two discrete cyclopropanone intermediates are possible; one leads to a linear beta-keto acid, which, through decarboxylation, yields octenones, while the other leads to a branched system capable, through enolization, of cyclizing to the gamma lactone. The conclusion from these studies must be that the reactivity of bromine in these acetate-derived compounds provides a mechanism for the synthesis of branched acyclic compounds of unusual types.

HALOGENATION AND REARRANGEMENT OF TERPENOID COMPOUNDS

Bromoterpene biosynthesis. In marine algae terpenoid compounds of all classifications (except the tetraterpenoids) are produced via halogen-induced biosynthesis. Based upon the structures of the compounds isolated, and biomimetic studies[2,3], such synthesis appears to consist of a Markownikov addition of a bromonium ion (Br^+) to an olefinic bond, followed by one of several possible reactions, such as chloride ion and water addition, proton elimination, and, most importantly, recyclization of the precursor to yield cyclic structures as shown in Figure 2. As this mechanism of cyclization closely parallels the classical proton-induced terpene cyclization found in terrestrial organisms, it generally results in brominated compounds of well known structure types.

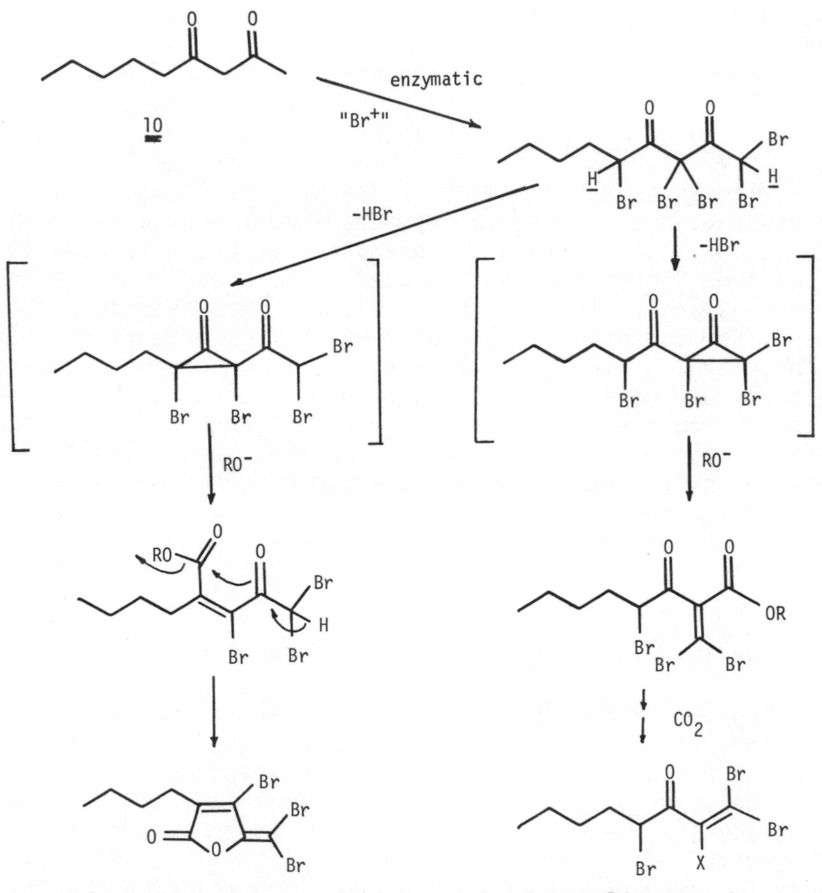

Figure 1. Proposed biogenesis for Delisea and Bonnemaisonia compounds involving Favorsky rearrangements.

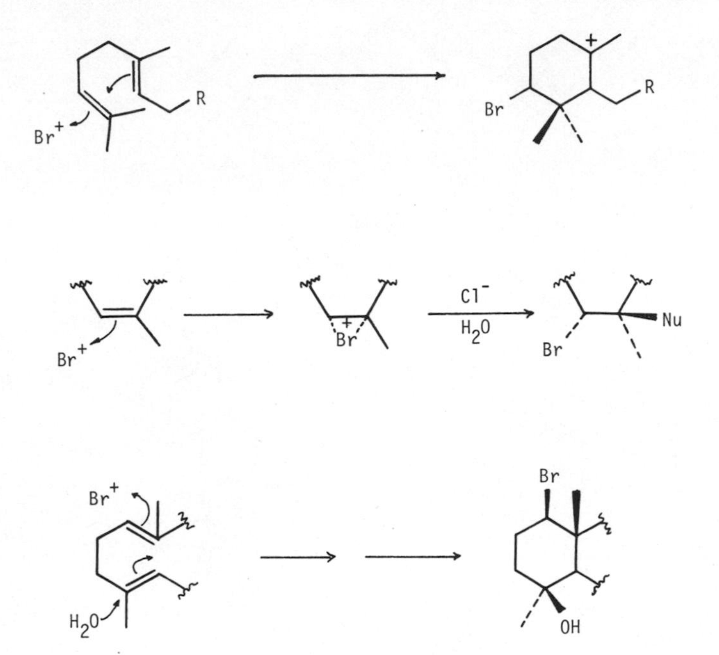

Figure 2. Bromonium ion-induced biosynthesis of marine-derived terpenoids.

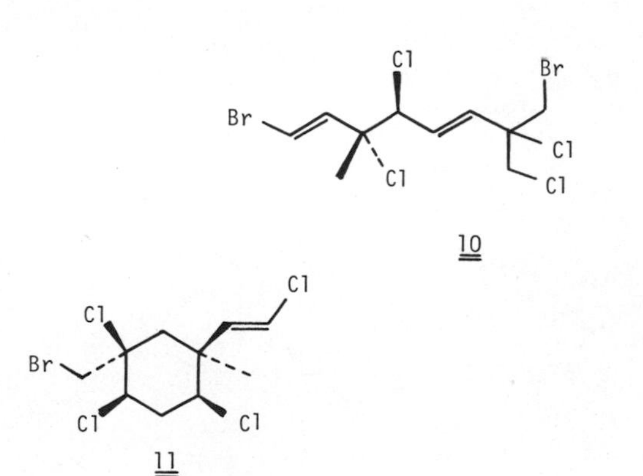

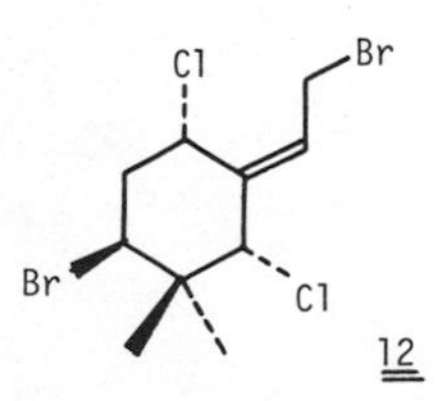

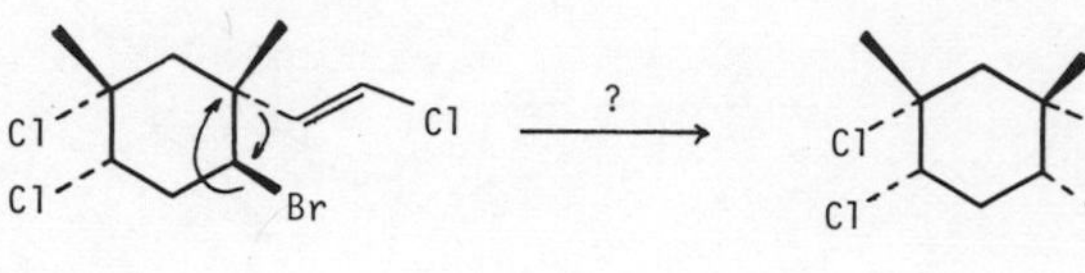

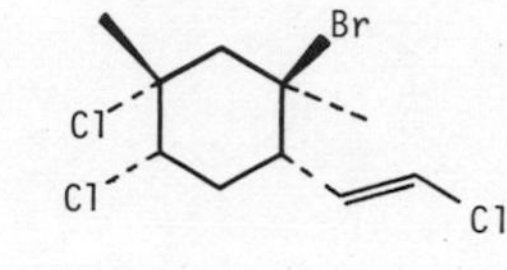

16 (not isolated) 13

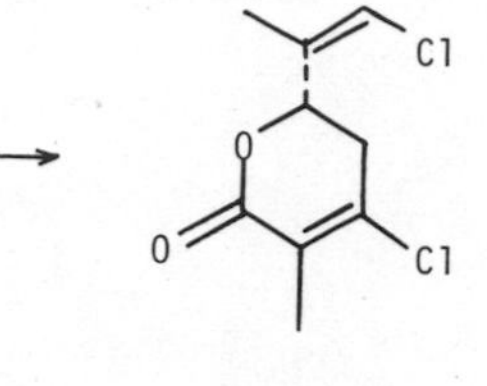

17 

Figure 3. Apparent biogenesis of monoterpenoid-derived marine natural products.

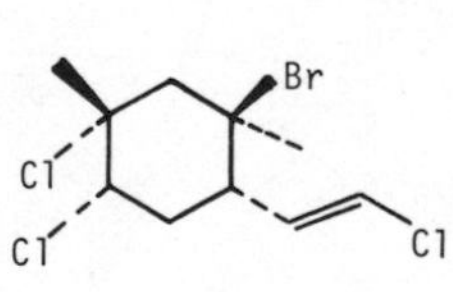

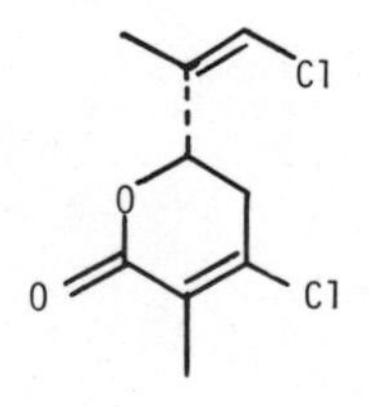

 14

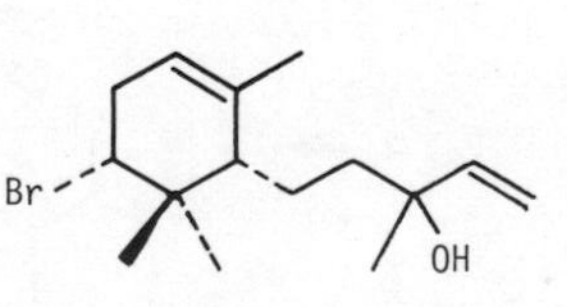

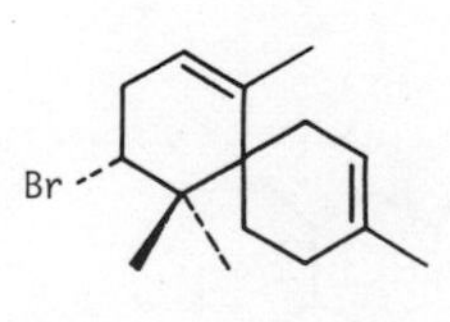

18 19

Monoterpenoids. Halogenated monoterpenoids have been isolated from only three genera of the red seaweeds, *Plocamium*[1], *Chondrococcus*[1], and *Ochtodes*[12]. Polyhalogenated linear compounds, such as 10, and cyclic structures such as 11 and 12, adequately show the "regular terpenoid" (normal head-to-tail isoprene linkage) structures of these compounds, which are produced by methods mentioned above. Two recently described compounds, the rearranged monoterpene 13 and the C_9 lactone 14, both from *Plocamium* spp.[13,14], are unprecedented monoterpenoid-derived compounds likely to have been produced *via* halogen-based rearrangements. While it has not been rigorously shown, the rearranged monoterpene 13 was proposed[13] to be derived from a potential precursor such as 16 by solvolytic removal of halogen and rearrangement, as shown in Figure 3. The probable origin of the degrated C_9 lactone was more securely predicted after the isolation and reactions of the likely *in vivo* precursor, 17, from the same source[14]. Treatment of the stable hemiketal 17 with diazabicyclo-undecane (DBU) in diethyl ether resulted in the apparent elimination of $CHBr_2^-$, and the quantitative production of 14, by a mechanism seemingly comparable to the iodoform reaction of methyl ketones. This reaction is clearly facilitated in 17 by the presence of the dibromomethyl group and the relative stability of the resultant $CHBr_2$ anion.

Sesquiterpenoids. The brominated, but regular, sesquiterpenoids α-snyderol (18)[15] and 10-bromo- α-chamigrene (19)[16] are classical examples of marine natural products whose analogous but unbrominated structure types are well known in terrestrial sources. A series of methyl-shifted compounds, 20 - 23, have recently been isolated from marine algae and their associated herbivores [17,18]. Based upon the aforementioned concepts, these compounds would appear to be derived from α-snyderol (18). This hypothesis requires a methyl migration to the developing carbonium ion center rather than several alternative reactions. To test the biomimetic propensity of this rearrangement, α-snyderol was treated with silver acetate at room temperature for 12 hours. The major product of this reaction was the rearranged acetate 24, in full support of the methyl migration hypothesis[19].

A similar series of speculations have recently been proposed to account for the rearranged skeletons of two compounds perforenone (25) and perforatone (26)[20], both iso-

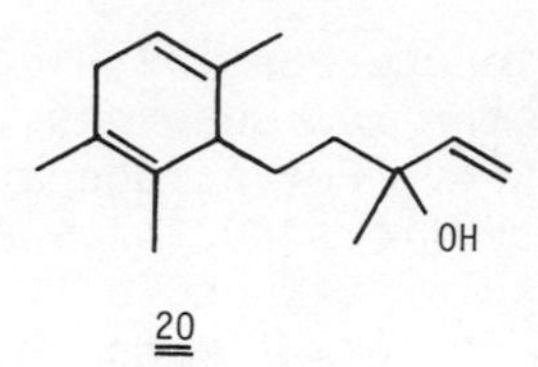

20

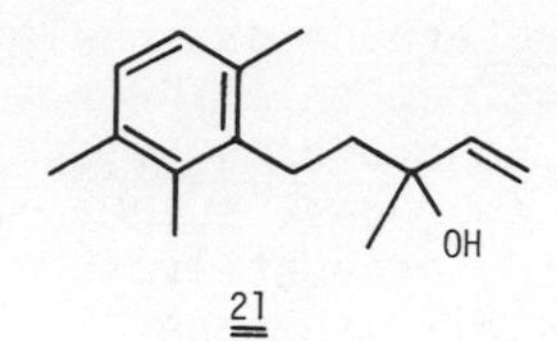

21

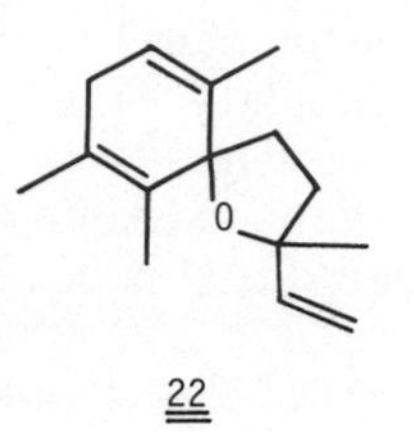

22

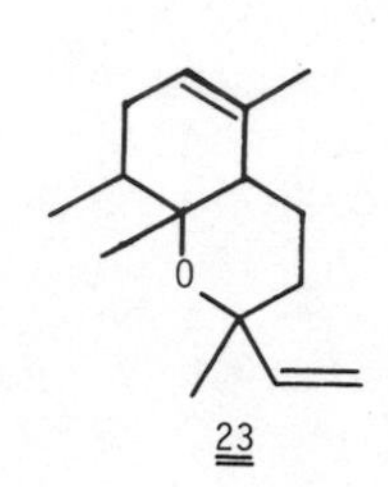

23

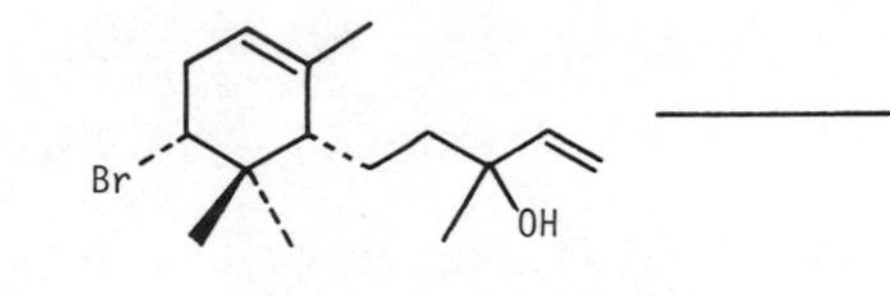

18

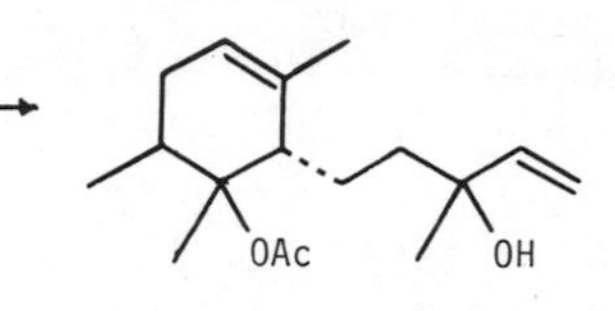

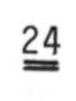

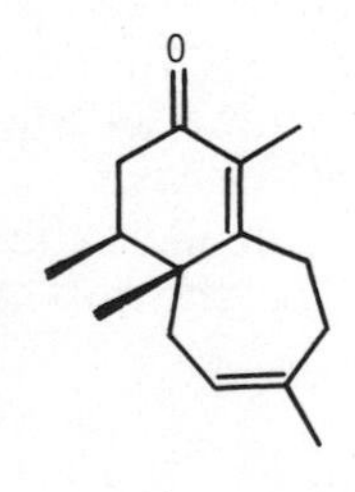

25

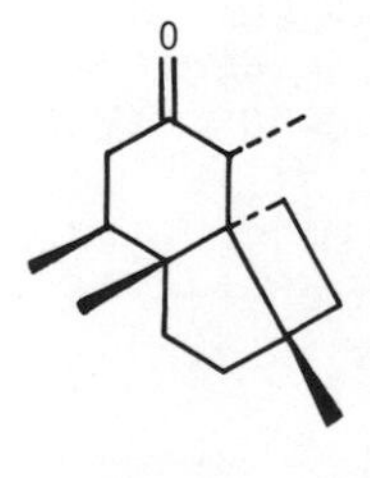

26

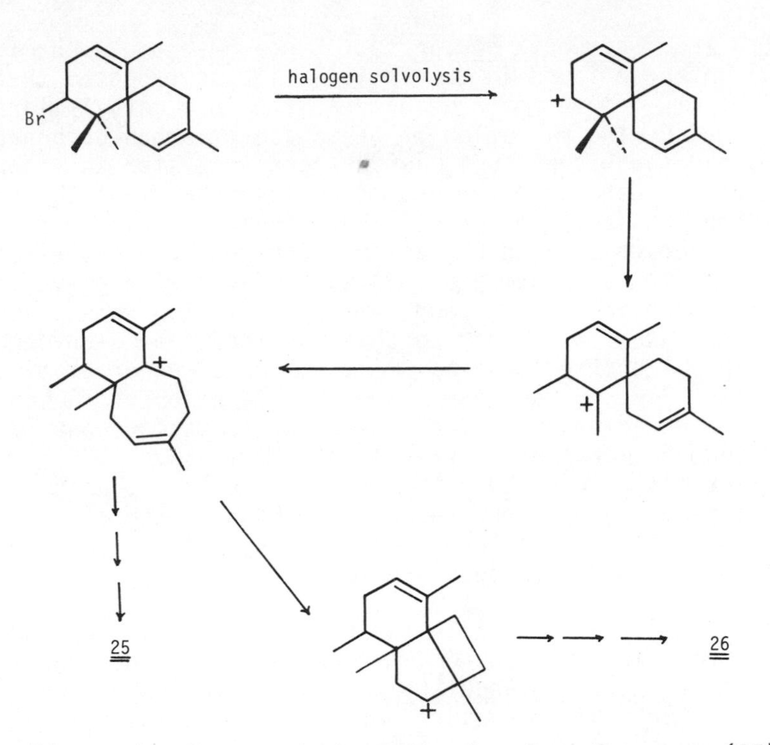

Figure 4. Proposed biosynthesis of perforenone (25) and perforatone (26).

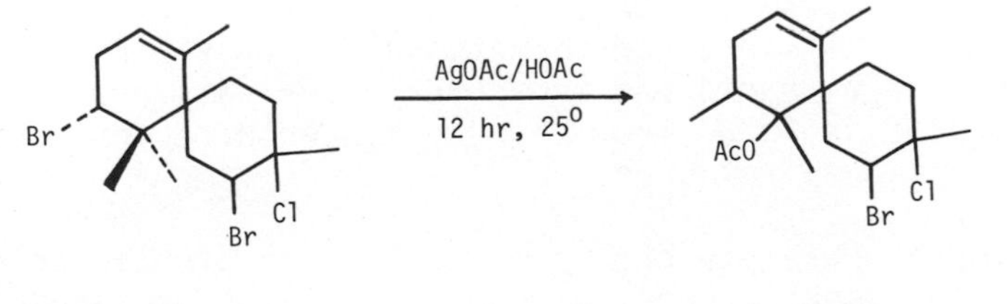

lated from Laurencia perforata. Since 10-bromo-α-chamigrene (19) has been found in Laurencia[16], it was proposed that solvolytic removal of bromine resulted in a methyl migration followed by the migration of an adjacent carbon-carbon bond to yield the bicyclo[5,4,0]undecane ring system as shown in Figure 4. Here also the proposed biosynthesis of 25 and 26 assumed the facile nature of the methyl migration. To assess the ease of this reaction with chamigrene derivatives, the trihalo chamigrene 27 was treated with silver acetate in acetic acid at room temperature. The sole product obtained was, in analogy to that obtained from α-snyderol, the methyl shifted acetate 28[19]. In more recent biomimetic studies[21a,b], it has been shown that debromoperfora-tone (29) is efficiently converted by base to perforenone (25), and that bromination of the alcohol derivative 30 (perforenol) yields 29. These stereospecific transformations, along with our recent isolation of the alcohol 30 as a natural product from Laurencia)[19], support these proposals as biologically feasible pathways.

One of the very early marine-derived sesquiterpenoids to be described was spirolaurenone (31) from the Japanese seaweed Laurencia glandulifera[22]. Spirolaurenone has a regular terpenoid structure but this ring system is unique to this one example. Since the only spirane system in Laurencia species has proven to be of the chamigrene type, it seems logical to consider spirolaurenone to have a similar origin. Recent work with an unrelated Laurencia metabolite, the diterpenoid obtusadiol (32), has, perhaps, provided evidence for another form of halogen-based ring contraction[23]. Obtusadiol possesses a rigid cis 1,2-bromohydrin configuration with an axial hydroxyl group, which when treated with dilute base undergoes a facile ring contraction to yield the methyl ketone 33. This reaction involves a formal elimination of HBr and appears to be facilitated by the rigid stereochemistry of the bromohydrin group (the equatorial Br facilitates a trans-anti carbon migration). While other mechanisms are possible, it seems likely that a chamigrene bromohydrin derivative such as 34 (as yet not isolated from Laurencia) could be the logical precursor to spirolaurenone.

Another example of the possible role of halogen elimination in sesquiterpene rearrangements lies, perhaps, with the bromoselinane derivative 35, isolated from an undes-

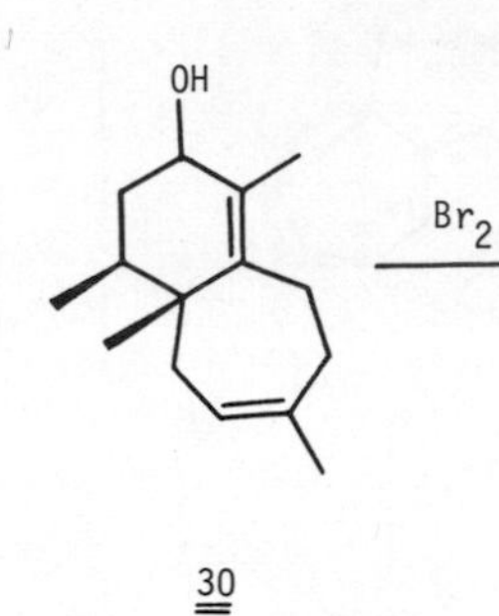

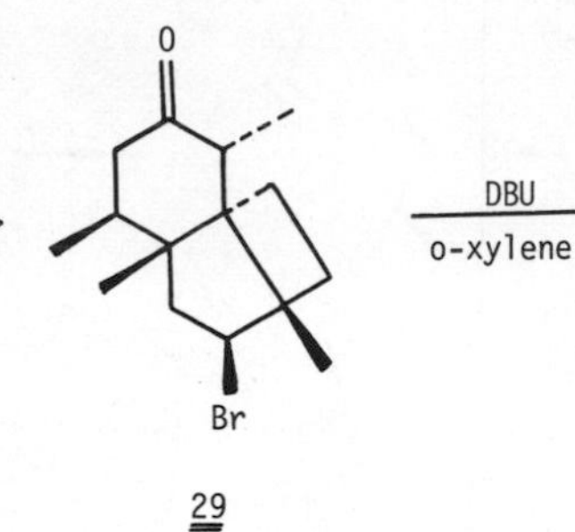

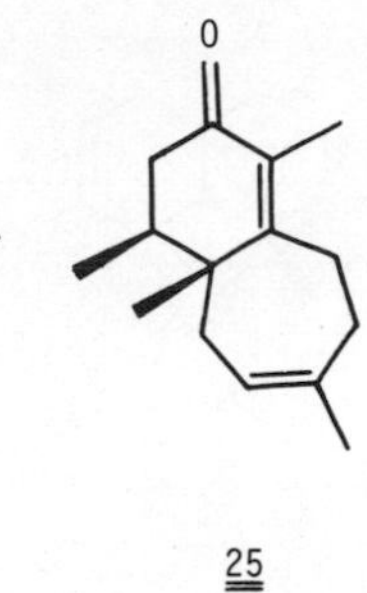

30 29 25

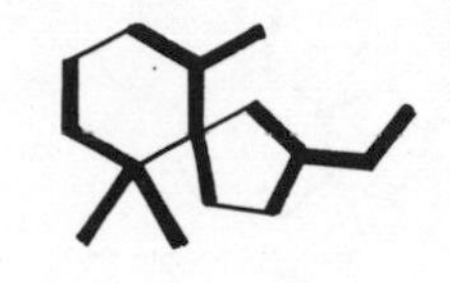

TERPENOID

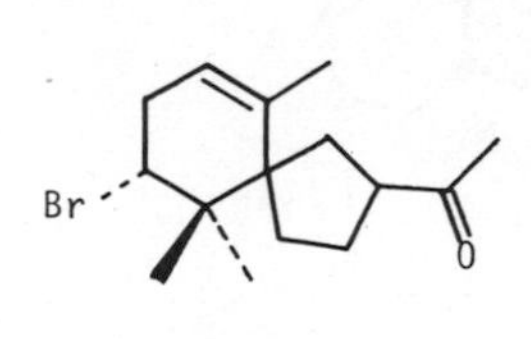

31

NONTERPENOID ?

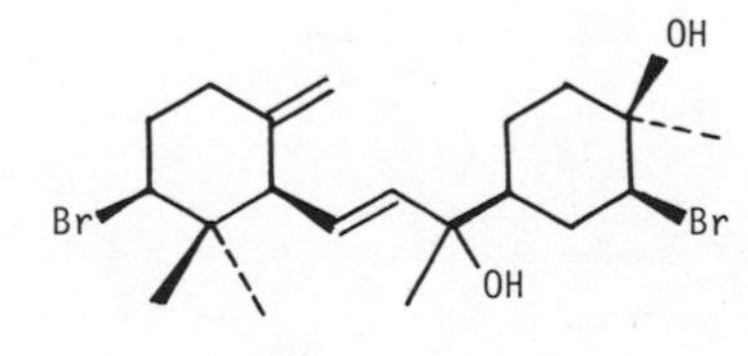

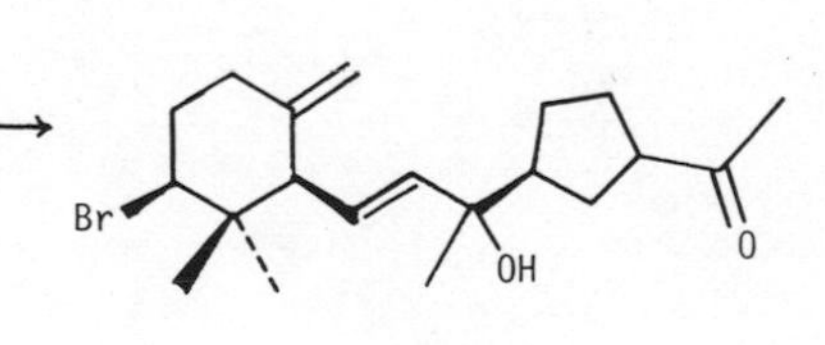

32 33

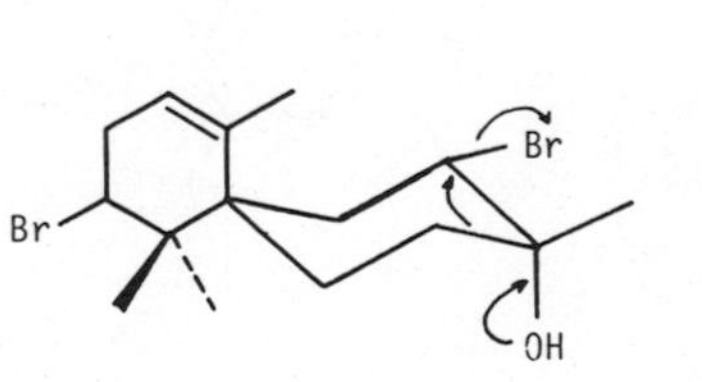

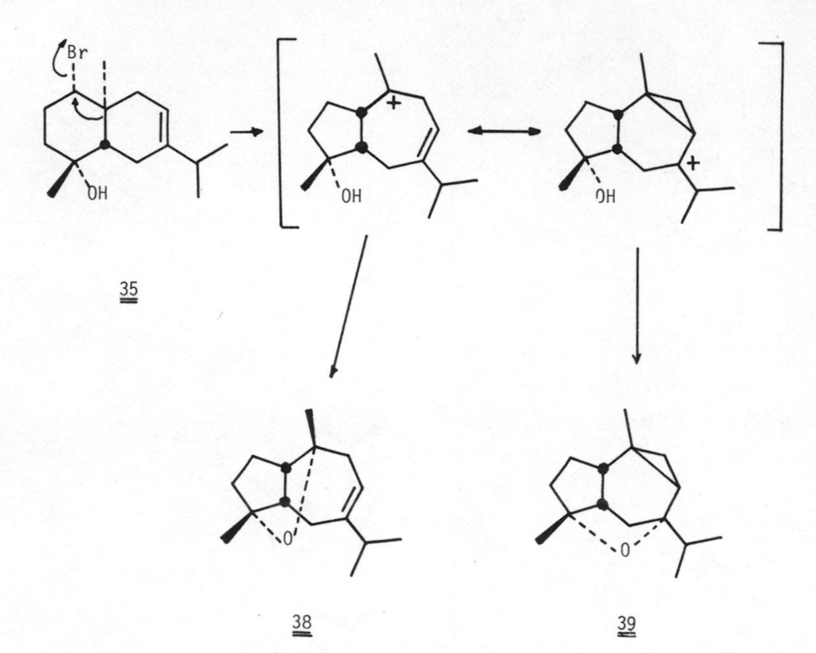

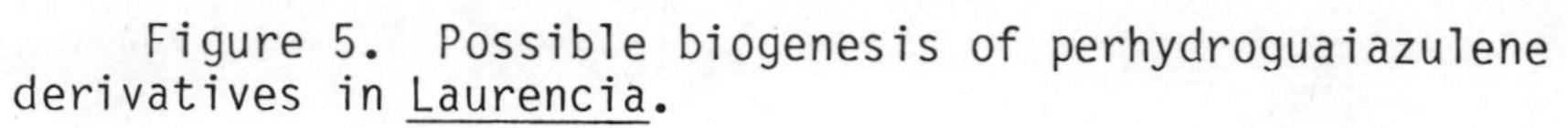

Figure 5. Possible biogenesis of perhydroguaiazulene derivatives in Laurencia.

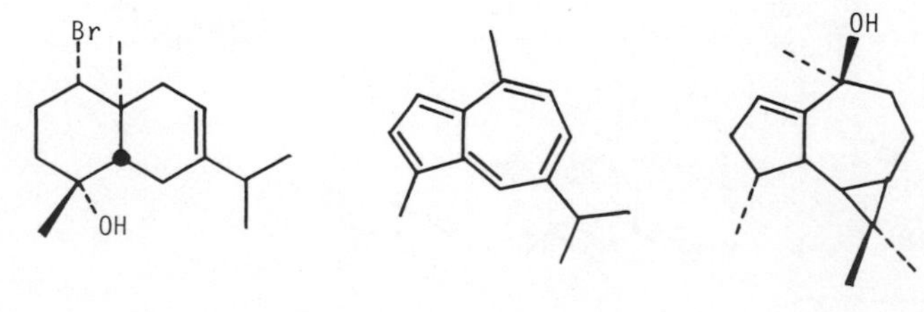

cribed species of Laurencia[24]. Detected along with 35 was the well known hydrocarbon guaiazulene (36), and it is relevant that perhydroguaiazulene derivatives such as 37 have been isolated from Laurencia[25]. The equatorial bromine substituent of 35 appears moderately stable toward solvolysis; however, when 35 is treated with silver acetate/acetic acid and warmed to 60°, a very clean conversion to the isomeric ethers 38 and 39 occurs. As illustrated in Figure 5, this conversion can readily be accounted for by the loss of bromine and migration of the bridgehead bond to yield the interconverting carbonium ion pair which is subsequently trapped by the hydroxyl function. The production of perydroguaiazulene derivatives in Laurencia by alternative methods such as the proton-induced cyclization of germacrene derivatives is, however, certainly not precluded.

Diterpenoids. Halogenated diterpenoids are more rare than sesquiterpenoids in the marine environment, and only two of these compounds are apparently products of halogen-based modification. Isolated along with the regular diterpenoid concinndiol (40) in Laurencia snyderiae[26] was the unique rearranged hydroperoxide 41. The identical structural elements of these compounds, and their simultaneous isolation, strongly argues that they are biogenetically related. Based upon this reasoning it has been proposed that bromine removal yields a ring-contraction reaction and subsequent oxidation occurs (Figure 6)[26]. It is interesting, however, that this system appears to yield a ring contraction whereas the identical bromo-cyclohexane system in snyderol gives methyl migrations.

The last diterpenoid rearrangement possibly induced as a consequence of bromine metabolism lies in the interesting diterpenoid sphaerococcenol A (42) from the Mediterranean red alga Sphaerococcus coronopifolius[27]. Sphaerococcenol A cannot be dissected into isoprenoid units, and after several possible biogenetic routes are considered, it appears that the most plausible precursor lies in the isoprenoid but irregular diterpenoid skeleton 43. The conversion of 43 to 42 requires a methyl migration, and based upon the generation of a unique bromomethyl group, the migration has been proposed, as illustrated in Figure 7, to involve the bromonium ion-induced opening of a cyclopropane-ring-containing intermediate.

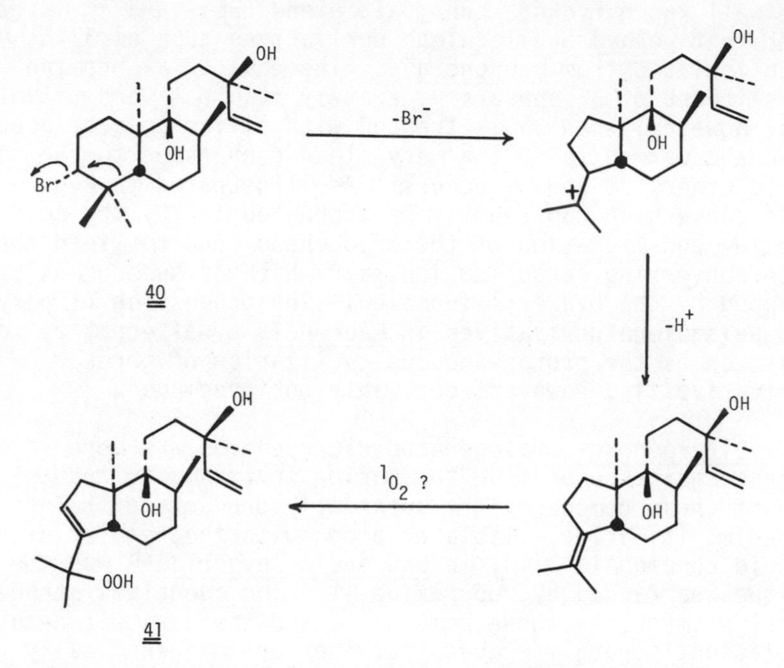

Figure 6. Proposed biogenesis of hydroperoxide 41 from concinndiol (40).

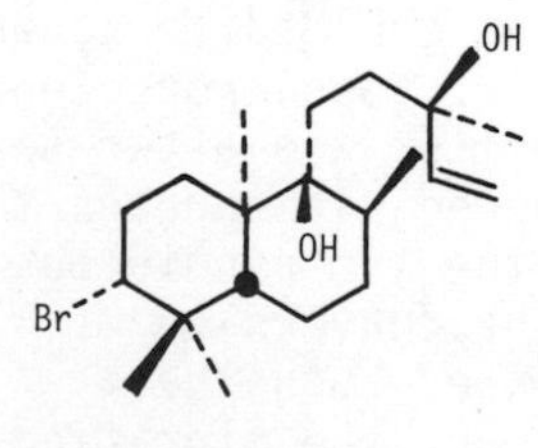

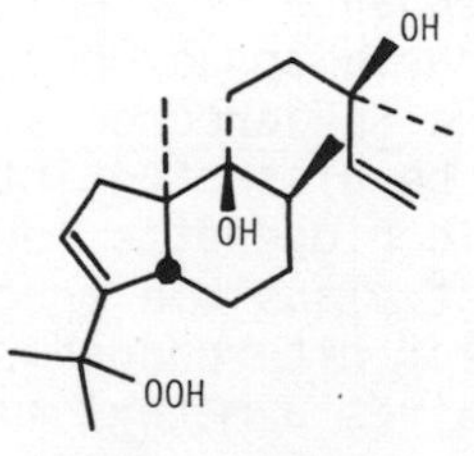

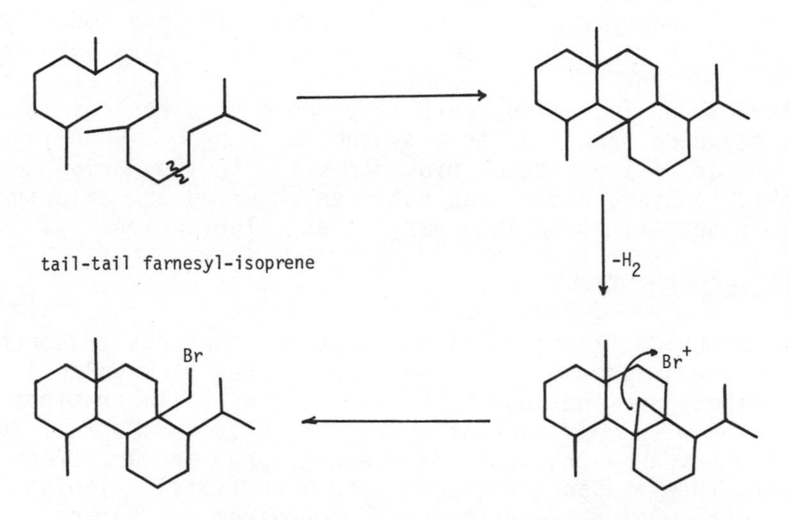

Figure 7. Potential biogenesis of sphaerococcenol A.

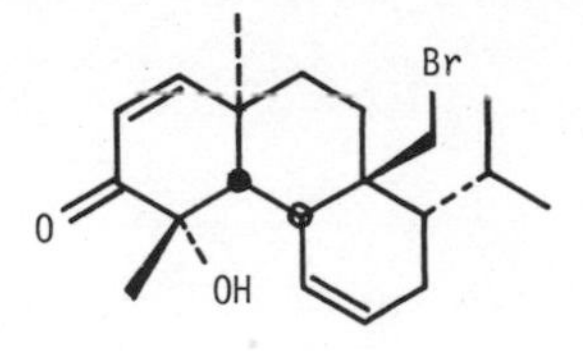

42

43

CONCLUSIONS

From the observations and concepts presented above, the reactivity of bromine, with respect to both addition reactions and subsequent elimination and substitution reactions, represents a major aspect of the production of marine-derived secondary metabolites. While chlorine and iodine are observed in marine-derived compounds, their roles in halogen-based synthesis are poorly understood. It is clear that chlorine and iodine do react to produce halo ketones in some red seaweeds, and that this synthesis is probably analogous to bromonium ion-induced biosynthesis. In the marine terpenes, however, iodine has not been observed and chlorine mainly appears to be incorporated as chloride ion.

ACKNOWLEDGEMENTS

I would like to point out that the concepts presented in this paper, as well as the experimental work which is described, are the result of the fruitful collaborations I have had with my former students, Dr. Oliver McConnell (now at the Skidaway Institute of Oceanography) and Dr. Bruce Howard (now at San Francisco State University). I wish to also acknowledge the generous support from the Marine Chemistry Program, Oceanography Section, National Science Foundation, under grant OCE 75-03824, which made our halogenation studies feasible.

REFERENCES

1. Fenical, W. 1975. Halogenation in the Rhodophyta; a review. J. Phycol. 11:245-259.
2. Faulkner, D. J. 1976. Biomimetic synthesis of marine natural products. Pure Appl. Chem. 48:25-28.
3. Wolinsky, L. E. and D. J. Faulkner. 1976. A biomimetic approach to the synthesis of Laurencia metabolites. Synthesis of 10-bromo-α-chamigrene. J. Org. Chem. 41:597-600.
4. Theiler, R. F., J. F. Siuda and L. P. Hager. 1978. Bromoperoxidase from the red alga Bonnemaisonia hamifera in Food and Drugs from the Sea, P. N. Kaul and C. J. Sindermann, Eds. University of Oklahoma Press, Norman, Oklahoma. pp. 153-169.
5. McConnell, O. J. and W. Fenical. 1976. Halogen chemistry of the red alga Asparagopsis. Phytochemistry 16:367-374.
6. McConnell, O. J. and W. Fenical. 1977. Halogenated metabolites--including Favorsky rearrangement products -- from the red seaweed Bonnemaisonia nootkana. Tetrahedron Lett. 1977:4159-4162.
7. Siuda, J. F., G. R. Van Blaricom, P. D. Shaw, R. D. Johnson, R. H. White, L. P. Hager and K. L. Rinehart, Jr. 1975. 1-Iodo-3,3-dibromo-2-heptanone, 1,1,3,3-tetrabromo-2-heptanone and related compounds from the red alga Bonnemaisonia hamifera. J. Amer. Chem. Soc. 197:937-938.
8. Geissman, T. A. and D. H. G. Crout. 1964. Organic Chemistry of Secondary Plant Metabolism. Freeman, Cooper and Co., San Francisco, pp. 70-72.
9. Pettus, J. A., Jr., R. M. Wing and J. J. Sims. 1977. Marine natural products XII. Isolation of a family of multihalogenated gamma-methylene lactones from the red seaweed Delisea fimbriata. Tetrahedron Lett. 1977:41-44.
10. McConnell, O. J. and W. Fenical. 1977. Polyhalogenated 1-octen-3-ones, antibacterial metabolites from the red seaweed Bonnemaisonia asparagoides. Tetrahedron Lett. 1977:1851-1854.
11. Kazlauskas, R. P. T. Murphy, R. J. Quinn, and R. J. Wells. 1977. A new class of halogenated lactones from the red alga Delisea fimbriata. Tetrahedron Lett. 1977:37-40.

12. McConnell, O. J. and W. Fenical. 1978. Ochtodene and ochtodiol, novel polyhalogenated monoterpenes from the red seaweed Ochtodes secondiramea. J. Org. Chem. in press.
13. Mynderse, J. S., D. J. Faulkner, J. Finer, and J. Clardy. 1975. (1R,2S,4S,5R)-1-Bromo-trans-2-chlorovinyl-4,5-dichloro-1,5-dimethylcyclohexane, a new monoterpene skeletal type from the red alga, Plocamium violaceum. Tetrahedron Lett. 1975:2175-2178.
14. Stierle, D. B., R. M. Wing and J. J. Sims. 1976. Marine natural products XI, Costatone and costatolide, a new halogenated monoterpenes from the red seaweed Plocamium costatus. Tetrahedron Lett. 1976:4455-4458.
15. Howard, B. M. and W. Fenical. 1976. α and β-Snyderol, new bromo-monocyclic sesquiterpenes from the seaweed Laurencia. Tetrahedron Lett. 1976:41-44.
16. Howard, B. M. and W. Fenical. 1976. 10-Bromo-α-chamigrene. Tetrahedron Lett. 1976:2519-2520.
17. Sun, H. H., S. M. Waraszkiewicz and K. L. Erickson. 1976. Sesquiterpenoid alcohols from the Hawaiian marine alga Laurencia nidifica III. Tetrahedron Lett. 1976:585-588.
18. Schmitz, F. J. and F. J. McDonald. 1974. Marine natural products, dactyloxene, a sesquiterpene ether from the sea hare Aplysia dactylomela. Tetrahedron Lett. 1974:2541-2544.
19. Howard, B. M. and W. Fenical. 1978. Unpublished results.
20. Gonzalez, A. G., J. M. Aguilar, J. D. Martin, and M. Norte. 1975. Sesquiterpenoids from the marine alga Laurencia perforata. Tetrahedron Lett. 1975:2499-2502.
21a. Gonzalez, A. G., J. Darias, and J. D. Martin. 1977. Biomimetic interconversion of two types of metabolite from Laurencia perforata. Tetrahedron Lett. 1977:3375-3378.
21b. Martin, J. D. and J. Darias. 1978. Algal sesquiterpenoids. Marine Natural Products, Chemical and Biological Perspectives, P. J. Scheuer, Ed. Academic Press.
22. Suzuki, M., E. Kurosawa and T. Irie. 1970. Spirolaurenone, a new sesquiterpenoid containing bromine from Laurencia glandulifera Kutzing. Tetrahedron Lett. 1970:4995-4998.

23. Howard, B. M. and W. Fenical. 1978. Obtusadiol, a unique dibromoditerpenoid from the marine alga Laurencia obtusa. Tetrahedron Lett. 1978:2453-2457.
24. Howard, B. M. and W. Fenical. 1977. Structure, chemistry and absolute chemistry of 1(S)-bromo-4-(R)-hydroxy-(-)-selin-7-ene from a marine red alga Laurencia sp. J. Org. Chem. 42:2518-2520.
25. Wratten, S. J. and D. J. Faulkner. 1977. Metabolites of the red alga Laurencia subopposita. J. Org. Chem. 42:3343-3349.
26. Howard, B. M., W. Fenical, J. Finer, K. Hirotsu, and J. Clardy. 1977. Neoconcinndiol hydroperoxide, a novel marine diterpenoid from the red alga Laurencia. J. Am. Chem. Soc. 99:6440-6441.
27. Fenical, W., J. Finer, and J. Clardy. 1976. Sphaerococcenol A; a new rearranged bromoditerpene from the red alga Sphaerococcus coronopifolius. Tetrahedron Lett. 1976:731-734.

INDEX